Design and Manufacture for
Sustainable Development 2004

Organizing Committee

Chairman
Tracy Bhamra

Co-Chairman
Bernard Hon

Members

Allen Clegg	Loughborough University
Richard Dodds	Liverpool University
Vicky Lofthouse	Loughborough University
George Moore	Liverpool University
Shahin Rahimifard	Loughborough University
Don Ritchie	Liverpool University

International Scientific Committee

L Alting	Technical University of Denmark, Denmark
S Evans	Cranfield University, UK
J Jesweit	Queen's University, Canada
F Jovane	Politecnico di Milano, Italy
H Kaebernick	University of New South Wales, Australia
F Kimura	University of Tokyo, Japan
C Luttropp	Royal Swedish Institute, Sweden
J McQuaid	Royal Academy of Engineering, UK
A Y C Nee	National University of Singapore, Singapore
C Ryan	Royal Melbourne Institute of Technology, Australia
J Seliger	Technical University of Berlin, Germany
M Simon	Sheffield Hallam University, UK
J A Stuart	Purdue University, USA
N P Suh	Massachusetts Institute of Technology, USA
H van Brussel	Katholieke Universiteit Leuven, Belgium
S Walker	University of Calgary, Canada
R Wertheim	ISCAR, Israel
W Wimmer	Technical University of Vienna, Austria
J Y Zhu	Nanjing University of Aeronautics and Astronautics, China
R Zust	Swiss Federal Institute of Technology, Switzerland

Design and Manufacture for Sustainable Development 2004

1st–2nd September 2004 at
Burleigh Court, Loughborough University, UK

Edited by
Tracy Bhamra
and
Bernard Hon

Organized by

Sponsored by

Sustainable Design Network

Professional Engineering Publishing Limited
Bury St Edmunds and London, UK

First Published 2004

This publication is copyright under the Berne Convention and the International Copyright Convention. All rights reserved. Apart from any fair dealing for the purpose of private study, research, criticism or review, as permitted under the Copyright, Designs and Patents Act, 1988, no part may be reproduced, stored in a retrieval system, or transmitted in any form or by any means, electronic, electrical, chemical, mechanical, photocopying, recording or otherwise, without the prior permission of the copyright owners. *Unlicensed multiple copying of the contents of this publication is illegal.* Inquiries should be addressed to: The Academic Director, Professional Engineering Publishing Limited, Northgate Avenue, Bury St. Edmunds, Suffolk, IP32 6BW, UK. Fax:+44 (0) 1284 705271.

© Professional Engineering Publishing Limited 2004, unless otherwise stated.

ISBN 1 86058 470 5

A CIP catalogue record for this book is available from the British Library.

Printed by The Cromwell Press, Trowbridge, Wiltshire, UK

The Publishers are not responsible for any statement made in this publication. Data, discussion, and conclusions developed by Authors are for information only and are not intended for use without independent substantiating investigation on the part of potential users. Opinions expressed are those of the Authors and are not necessarily those of the Institution of Mechanical Engineers or its Publishers.

Preface

Sustainable Development is becoming a more pressing issue for all types of organizations and institutions. Whether driven by legislation, the 'greening of the marketplace', supply chain requirements, or the pressure of events associated with climate change, sustainable development is now becoming a matter that must be addressed at both a strategic and operational level. However, the issues are complex and the supporting structures are weak, so there is much to be done in order to enable those who must take sustainable development into account when making decisions to do so effectively. It is for that reason that effort needs to be focused in changing the education of designers and engineers, and also in developing useful methods and tools to enable sustainable development thinking to be part of all organizational practices. These subjects are still in their infancy and designers and engineers still need to gain understanding of the subject and support in tackling the issues it highlights.

As researchers and practitioners gain experience of sustainable development, more information is emerging. This volume is based on the papers presented at the 3rd International Conference on Design & Manufacture for Sustainable Development, held at Loughborough University from 1st–2nd September 2004 and highlights the way that the subject is developing and provides a useful insight into the way design and manufacturing are changing. It covers subjects such as Sustainable Design Education, Strategies for End of Life, Understanding Consumers to Create Sustainable Products, Methods and Tools for Sustainable Design, and Case Studies of Sustainable Design.

About the Editors

Tracy Bhamra has been a Senior Lecturer in the Department of Design and Technology at Loughborough University since April 2003. Prior to this she spent over six years at Cranfield University where she established the MSc in Manufacturing: Sustainability and Design, the UK's first postgraduate course focusing on Sustainable Design. Her research interests are focused around Sustainable Design and its use and implementation across all industrial sectors. She has held a number of UK Government research grants for projects investigating Sustainable Design and its implementation in industry. Dr Bhamra also runs the Sustainable Design Network in the UK, with over 160 members from Industry, Academia, and Public Bodies, which is aiming to develop partnerships to encourage the implementation of Sustainable Design. She is a Chartered Engineer, a member of the IEE, and a Fellow of the RSA.

Bernard Hon is the holder of the Chair of Manufacturing Systems at Liverpool University since it was first established in 1990. His research interests are focused on new manufacturing technology, especially rapid and micro manufacturing processes, design of manufacturing systems, and more recently, design and manufacture for sustainable development. Professor Hon is also the Programme Director of two MSc courses in Advanced Manufacturing Technology and Product Design. Professor Hon has served on numerous national and international bodies such as the British Council, Croucher Foundation, IEE, Foresight, and has acted as advisor in manufacturing systems to British and overseas universities and government agencies. He is an active member of CIRP and a Fellow of the IEE.

Contents

Sustainable Design Education

Methods and Tools for Sustainable Design

Case Studies of Sustainable Design

Keynote Papers

Object lessons – enduring artefacts and sustainable solutions

S WALKER
Faculty of Environmental Design, University Calgary, Canada

[Handwritten annotations:]

1. shift Δ in society values >> NPD

2. - time dependent — beyond functional necessary
— Δ fashion
— style/fashion features of developed economy where basic needs are satisfied.

ABSTRACT

This paper explores sustainable product design by considering selected artefacts that have been produced and used in human societies for thousands of years. These are categorised according to their dominant characteristics as *functional*, *social/positional*, and *inspirational/spiritual*. Products that combine these categories are also considered, and one object, that combines all three, is discussed in more detail. Lessons are drawn for our current understandings of sustainable products. Examples of contemporary products are also discussed and some experimental designs by the author are included which attempt to balance product characteristics in ways that also address sustainable principles.

3. Product too 'inexpensive'

1 INTRODUCTION

The relationship between sustainability and product design has been the subject of extensive debate in recent years and is clearly complex and multifaceted. It is frequently presented in terms of product life cycle, materials, manufacturing, and environmental issues (1,2), there has also been considerable discussion about the design of longer-lasting products and of the links between products and services (3,4). These approaches make important contributions to sustainability, they can help reduce product impacts and improve production efficiencies. However, they also tend to be rather prosaic and dominated by pragmatic concerns. As such, they seldom ask more fundamental questions about the meaning and place of products in our lives, and the contribution of material goods to what might be broadly termed 'the human endeavour'. Therefore, these approaches neither address the crux of the problem nor do they allow us to fully appreciate the magnitude of the shift in attitudes and expectations that is demanded by 'sustainability'.

Here, a different path is taken in an attempt to address this deficiency and to, hopefully, further our understanding of sustainability and product design. Artefacts are considered in terms of their characteristics and meanings. The artefacts have been specifically chosen because they have existed in one form or another in human societies for millennia, and are still made and used today. When objects have been produced over such long periods of time, spanning diverse cultures, languages, and understandings, then we can be sure that there are lessons to be learned from them about our relationships with material things, and our contemporary efforts to tackle sustainable issues in product design and manufacturing.

In pursuing this line of thought, objects have been classified into three broad categories: 1) functional, 2) social/positional, and 3) inspirational/spiritual. The characteristics of objects in each of these categories are described and their relationship to sustainability is discussed. These broad areas overlap and objects that combine these characteristics are also discussed; and it is suggested that the *'functional'* plus *'social /positional'* combination is the most problematic in terms of sustainability. Furthermore, there are some objects that combine all three classifications. One of these in particular has been present in human societies for thousands of years and is ubiquitous today across much of the world. This object has an exceptionally intense sense of "possession-ness" associated with it. For these reasons, this object will be considered in rather more detail to explore what lessons it might hold for sustainable design and manufacturing.

2 SUSTAINABLE OBJECTS

There are numerous examples of artefacts that have been produced since very early times and are still in common use today. A perusal of the collections in many of our large national museums (5) reveals that certain kinds of objects have been prevalent in human society since very early times; notably items such as pottery, tools, weapons, jewellery, and statuary. These types of objects have been in continuous production for at least five thousand years, and the earliest examples of jewellery were recently estimated to be some 75,000 years old (6). These objects are generally valued for their utility, their decorative and aesthetics qualities, and/or their symbolic or ritualistic roles. The value attributed to an object will usually emphasise one of these over and above the others; and while a particular object might be rather ephemeral in terms of its materials, its style or its motif, the general 'object types' mentioned above have persisted over very long reaches of time.

These kinds of objects can rightly be characterised as *sustainable*; the sheer longevity of their production and use, and their continuous place in human society clearly testify to their enduring importance in supporting human existence or in nourishing human culture. Therefore, it will be useful to examine some of the general characteristics of these objects. In turn, these characteristics can be considered in relation to human needs and values, and can therefore inform our contemporary response to product design and sustainability.

3 OBJECT CHARACTERISTICS

The examples of enduring objects introduced above can be classified into three broad categories 1) *Functional*, 2) *Social/Positional*, and 3) *Inspirational/Spiritual*, see table 1. Each of these discrete categories emphasises quite different product types and characteristics:

Functional Objects: Tools, weapons and everyday pottery are valued primarily for their usefulness. If the tool is ineffective then its value is severely diminished – it would be described as 'useless'. Similarly, a weapon is judged by its usefulness in hunting or in affording protection,

D004/017/2004

and a ceramic pot by its ability to hold liquid. These objects are designed to accomplish practical tasks; they are intended primarily to be useful rather than attractive. Design considerations focus on effectiveness, safety and user comprehension. Therefore, their chief characteristic is *functionality*, see fig.1.

Social/Positional Objects: Jewellery items such as necklaces, earrings and bracelets, products such as make-up, cosmetics and tattoos, and badges, brooches and medals are all non-utilitarian. While they serve a purpose, they are not practical implements or utensils. Instead, they are used to express identity, to be decorative, to enhance one's appearance and one's attractiveness, or to indicate one's social standing, rank, achievement or affiliation. The chief characteristics of these products are their *social* or *positional* qualities (7,8). They serve as social signifiers that can enhance one's sense of self-esteem, one's social acceptance, or indicate one's social status, see fig.1.

Inspirational/Spiritual Objects: A third category can be classified as inspirational or spiritual in character; it includes religious statuary, icons, and fine art objects. These objects refer to or convey inspiring, higher, s acred or spiritual ideas. They are physical expressions of profound understandings and beliefs and because of this they are considered deeply meaningful objects. They often have religious, magical or talismanic associations and can serve as reminders or touchstones for our most deeply felt yearnings, see fig.1.

These three categories represent three very significant types of objects – objects that have stood the test of time and have held their place in human society irrespective of culture, class, beliefs and language. We can therefore conclude that such objects are "non-trivial" and, at least in terms of their continuous presence and use in human society, sustainable. We can infer that such objects fulfil important human needs. Indeed, when we compare the characteristics of these objects with our understandings of human needs, such as the modified version of Maslow's Hierarchy of Human Needs (9) and Hick's natural, ethical and spiritual meaning (10,11), it becomes clear that, taken together, these three sets of product characteristics correspond to a broad and comprehensive range of human needs:

Functional objects allow us to fulfil our physiological and biological needs as well as our safety needs, such as ensuring personal security or fending off danger.

Social/Positional objects including jewellery, make-up, medals and badges of identity respond to our need for love, belonging, social acceptance, our standing within a social group, our sense of achievement, and self-esteem.

Inspirational/Spiritual objects, such as religious artefacts or fine art pieces respond to our need to know, our search for meaning, our aesthetic sensibilities, personal growth, our spiritual needs, and our need to reach out beyond ourselves to help others attain their potential (12).

In addition to these, however, there are many objects that are not adequately described by just one of these three sets of characteristics; instead, they bridge two or, in some cases, all three of the categories. A consideration of these more complex objects yields, on the one hand, insights about those object types that are problematic in terms of sustainability, and on the other hand, object types that hold fundamental lessons for the design and manufacture of sustainable products.

We can identify objects that have both *functional* and *social/positional* qualities, others that have *spiritual/inspirational* and *social/positional* characteristics, and still others that have *functional, social/positional* and *spiritual/inspirational* characteristics. Objects that have only *functional* and *spiritual/inspirational* characteristics are probably not feasible (13). Let us now briefly look at objects that combine these various characteristics:

'Inspirational/Spiritual + Social/Positional' Objects include ornaments, commercial art pieces, souvenirs, and home décor items, see fig.2. This category also includes statuary or art objects that have social/positional meaning attributed to them, such as status, esteem or personal identity. It can also include items based on traditional cultures and religions such as the commercially produced Haida Masks of the Canadian west coast. These types of sculptures are produced today for the tourist or collector markets and in the process changes occur. Some of these changes can be very positive, creating new opportunities for artistic expression while simultaneously opening up new avenues for economic development and self-determination. However, the changes can also be negative. The objects can become modified, clichéd and stereotyped in order to serve the market (14,15). When these non-functional objects become commercialised, their religious, ritualistic or cultural significance is no longer relevant, they become primarily decorative and there is a danger of them becoming a pastiche or falling into kitsch.

These types of objects do not pose too much of a problem in terms of sustainability – on the contrary, their production can be a positive development. They are generally 'low-tech', being frequently handmade at the local level and employing local skills, cultural and aesthetic sensibilities and perpetuating cultural ties, albeit in some cases in a new and often diluted form. Taken to extremes this last point can become destructive to a culture's heritage. Nevertheless, local employment, use of local materials and local designs, frequently with natural materials, can be socially and economically beneficial and environmentally of relatively low impact. In addition, the handmade and cultural or personal significance of these types of objects means that people will often keep them for a long time – even passing them down from one generation to another. They are often regarded as precious personal possessions and they may have heritage value, which in turn prevents the object entering the waste stream.

'Functional + Social/Positional' Objects include functional consumer goods that have both utilitarian value and social, positional, or identity value, see fig.2. Products such as automobiles, watches, music equipment, footwear and 'designer-labelled' goods all possess positional value in addition to their essential utility. (Ornaments and souvenirs derived from functional objects such as decorative pots can also combine functionality with social value. However, in these cases the primary purpose is often decorative, and the functionality becomes largely irrelevant.)

These are functional products that set one apart from the crowd and in terms of sustainability they are, by far, the most problematic. For the most part they are mass-produced goods that are promoted and distributed globally. These goods drive consumerism and are the cause of many environmental and social ills. They not only combine functionality with positional value, they also become quickly outdated. Their significance and value is exceptionally time dependent. This

D004/017/2004

is because firstly, both their functionality and their positional value are intimately connected to advances in technology, and secondly, their positional value is tied to changes in fashion and styling. Within our contemporary market-driven, mass-production system, the linking of technological progress and/or styling with social status has become an extremely potent combination. Today virtually all our utilitarian goods have the potential to be positional, from cars and audio products to refrigerators, kettles and bathtubs. When this occurs an object's value is determined not simply by its ability to properly function, but also by its ability to convey social position, aspiration or affiliation. This positional value is inevitably short-lived because technology is always advancing and styling is always changing. It is these factors that spawn the upward spiral of consumerism that is so environmentally and socially destructive.

'Functional + Social/Positional + Inspirational/Spiritual' Objects are the final category to be considered here. This category includes objects related to religion and particularly to forms of prayer, for example a Muslim prayer mat, a Buddhist prayer wheel or a Jewish prayer shawl, see fig.3. Each serves a functional purpose, the prayer mat defines a space for prayer, each rotation of the prayer wheel represents a prayer's recitation, and the prayer shawl is a mnemonic device (16). Inseparable from these functions, each has a symbolic religious or spiritual significance, and each is a signifier of social identity and, potentially, each may also be associated with social status or position. These are important religious and cultural artefacts that all pertain to our inspirational or spiritual understandings, and each is 'used' in an active, functional way that is quite different from a religious statue or painting.

These types of artefacts are considered precious because of their sacred associations. Their design and use is steeped in tradition and they are not simply discarded when a newer model or style comes along. These objects can therefore be described as 'sustainable'. They have a long history in human society, they are highly valued, and they have profound meanings. That said, these examples, the prayer shawl, prayer wheel, and prayer mat, are each specific to a particular religious culture. There is, however, a similar object that is found all over the world and in most of the major religions. We will consider this object in rather more detail because it holds important lessons for our understandings of sustainability and material culture.

4 AN ENDURING OBJECT

Imagine an object that is used today by rich and poor, young and old, healthy and sick; an object that has a prosaic, utilitarian function and a deeply spiritual significance; that can be decorative and highly aesthetic; and has for its owner a profoundly personal value which is independent of price, quality or materials but is inherent to that *particular* object. Imagine, too, that such an object has a wide variety of designs and manifestations; that it can be mass-produced for a few pennies or, for a similar cost, made at home. Perhaps the contemplation of such an object would allow us to see anew some of the failings of our contemporary, rather limited approaches to product design and production, and offer some pointers for a more sustainable and more inclusive future.

In the tragedy of Baghdad a man scarred with the wounds of conflict holds this object (17). High in the Himalayas a young boy uses it to keep a tally. A smaller version can be seen in the fingers of an old man in a café in Athens. In New York it may be found in the pocket of a business suit or in a fashionable Gucci handbag. In many a Chinatown a stall can be found bursting with different versions in all shapes, sizes and colours. It is an object that crosses boundaries of time, belief, gender, culture and class. The year October 2002 – October 2003 was dedicated to it (18). In December 2003 five hundred of these objects were used by British artist Mark Wallinger to decorate the Christmas tree at Tate Modern in London (19). It is variously known as the 'mala', the 'tasbih', the 'rosary' or simply as 'prayer-beads' and through the centuries it has carved out a unique place in human culture as an object that ties the physical or outer person with the inner, contemplative and spiritual self.

The widespread and enduring use of prayer-beads, together with their fundamental relationship to the human search for meaning, make them an important artefact for consideration by the product designer seeking to better understand the relationship between sustainability and material things.

5 PRAYER-BEADS

> And if I bidde any bedes, but if it be in wrather,
> That I telle with my tonge is two myle fro myn herte.
> William Langland, 14th century England (20)

At their most basic functional level prayer-beads are used for keeping track of repeated chants or prayers. Their most common form is that of a simple circle of beads or knots on a string, ending in a tassel or religious symbol. They are thought to have originated in Hinduism about 3000 years ago (21). Buddhists have used the *mala* since very early times (22), the Muslim *tasbih* dates back to about the 9th century (23), and the Catholic *rosary* has been used since the 15th century (24,25). The Eastern Orthodox churches use knotted *prayer ropes*, an Anglican rosary was introduced in the 1980s (26), the Baha'i faith uses simple strings of beads similar to the *tasbih*, and there are also secular versions – in Greece small strings of 'worry-beads' are used to relieve stress, see fig.4.

5.1 Characteristics of Prayer-Beads
Let us look in a little more detail at the various uses and meanings of prayer-beads in order to develop some insights that will be useful to our understandings of product design and sustainability. The various uses and meanings of prayer-beads include:

A Tallying Device: A bead, representing one prayer in the cycle, is held in the fingers whilst the prayer is recited. In this respect, they serve as a functional, utilitarian tool.

An Aid to Concentration and Meditation: Essentially, prayer-beads are a device to assist concentration while praying a nd m editation (27,28). The f ingering o f t he b eads i s a r epetitive activity that can be done without thinking; importantly it is an activity that occupies the physical body. Pascal talked of using such routines in order to enable us to act unthinkingly and mechanically, in order to subdue the machine and the power of reason, (29,30). This is a critical aspect of prayer-beads; the repetitive action produces a quieting effect (31). We see similar mechanical routines practiced all over the world because they are associated with spiritual growth. The spinning of the prayer wheel, the raking of a Zen garden (32), or the rocking action of orthodox Jews during prayer (33). These practices are thought-less or 'unreasoned' actions, which facilitate meditation and, potentially, inner growth. It is this fundamental purpose that raises prayer-beads above the merely mundane and functional. The simple string of beads is an instrument of synthesis - an aid in bringing together the inner and outer, or physical and spiritual (34,35). Thus, the prayer-beads are profoundly meaningful, which, as we shall see, is relevant to our understanding of sustainability.

All the major spiritual traditions are expressed, on the one hand, through teachings and traditions that are often somewhat esoteric and difficult to grasp and, on the other hand, through popular understandings and customs. In this respect, prayer-beads have various other meanings that add to their widespread appeal.

A Talisman: Prayer-beads are often regarded as a lucky charm that wards off danger (36). In some religions losing one's prayer-beads is an ominous sign (37,38) and in Catholicism, even in recent times, the rosary has been associated with apparitions and miracles, it is commonly viewed as an object of comfort (39), and in many Latin countries it is a ubiquitous adornment of a car's rear-view mirror. These types of associations are deeply rooted in the human psyche and, despite scientific and technological progress and our rationalistic outlooks, they are still very much present in modern, secular societies. Other common examples include: the omission of row 13 in aircraft by major airlines in the world's most scientifically and economically advanced countries (40); and the commonly held superstition that walking under a ladder brings bad luck.

A Touchstone: Prayer-beads can serve as a 'remembering object'. It is not a mnemonic device in the usual sense. Rather, it serves as a benchmarking device, a 'reminder object', similar to a souvenir, but for a person of faith it is a reminder of values, of that which is true and meaningful in his or her life.

Jewellery: Prayer-beads can also be worn as jewellery, as a necklace or bracelet. In this case prayer-beads are valued for their aesthetic and decorative qualities.

A Badge of Identity: In various ways throughout their history, prayer-beads have been used as an outer sign of identity – an indication of one's religion, denomination or vocation (41).

To this point we have discussed the object in terms of its use and meanings. It can also be considered in terms of its physicality and materiality:

A Physical Expression of the Accompanying Prayer Cycle: In Catholicism the name *rosary*, is actually the same as the name of the prayers that accompany its use. The design of the rosary, a circlet of beads attached to a pendant with a crucifix, is essentially a tactile map and visual diagram of the prayer cycle. Hence, its physical design is an indicator of its use and meaning.

The Physical Qualities of the Object include the size, weight, colour and texture of the beads, whether they are warm or cold to the touch, and how they sound when they are picked up and used. These are key aspects of one's aesthetic experience of the object. They can be of plain wood or of precious jewels, simple or elaborate. The reasons for such variety can range from a genuine attempt to achieve an appropriate expression for a devotional object, to a choice that has more to do with social standing or (perhaps disingenuous) outward expressions of piety. Simple wooden beads can be an authentic expression of simplicity and humility, or a conscious attempt to indicate to others something of one's piety. A costly, bejewelled set of beads can be an entirely appropriate object for use in religious practice, or it can be a sign of wealth and social standing (42). Thus, the appearance of prayer-beads can be diverse, variously interpreted, and used to express a broad range of values.

Varieties of manufacture: Prayer-beads can be handmade from the simplest of materials or batch produced in larger numbers by local artisans. They are also commonly made by mass-production processes. How it is made, what it is made from, and where it is made may have a bearing on the value ascribed to it by its owner. However, a cheap, mass-produced set of beads can be as precious to its owner as a set made from rare and expensive materials. Moreover, prayer-beads often include an emblem identifying the place it was purchased, such as a pilgrimage site. This adds a souvenir quality to it, but also a particular sacred association.

From this brief overview it is apparent that there are a wide range of meanings associated with this object. They span the utilitarian, the deeply reflective and contemplative, the talismanic, the emblematic and the decorative. It can also serve as a touchstone of values and an indicator of social status. For these reasons, this object can acquire *an exceptionally intense, and highly personal quality of 'possession-ness'*, perhaps even more than the wedding ring; it is an object that one tends to really 'own' and cherish in a very intimate way (43) regardless of the fact that it may have cost very little and be made from mass-produced plastics.

There are two more aspects of prayer-beads that are important to bear in mind when considering the relationship between sustainability and the design of material objects:

Evolution over time: neither the 'prayer-beads' as an artefact, nor the cycle of sayings that accompanies its use were 'designed' as such. Rather, both evolved over a long period into the forms we see today. These forms a re the result of b oth popular (or bottom-up) practices and institutional (or top-down) approval and modification (44).

Evolution among different traditions: the different forms of prayer-beads around the world demonstrate that it is an object that is easily adaptable to diverse cultures and traditions, which then make it their own through modifying the design, and in doing so it becomes a symbol of both belief and identity. Hence, its flexibility allows it to become acculturated and this contributes to its continued but diverse use and meaning.

6 OBJECT LESSONS FOR SUSTAINABILITY

In this paper we have looked at various types of enduring objects, categorised them and discussed these various categories in terms of human needs and values. One object in particular, the prayer beads, has been discussed in rather more detail as a important example that spans the various categories that have been introduced. We can now look to the lessons this object might hold for sustainable product design, bearing in mind that we cannot necessarily draw any firm, generally

D004/017/2004

applicable conclusions from the specific characteristics of one object. Nevertheless, from the above discussion we can make the following observations:

The Physical and the Meaningful: Firstly, it seems that a very powerful sense of personal possession can be attributed to an artefact in which there are strong, interwoven relationship between physical object, physical activity, tactility, visual understanding, aesthetic experience, meaning, inner growth, and allusions to the numinous. Secondly, the object discussed here is fundamentally profound in its conception as a thing, and this is articulated through its physical design, its use and its meaning to its owner. It is a deeply evocative artefact that is neither trivial nor trendy, nor is it based on transient technological novelty or styling. For these reasons it is not susceptible to many of the factors that render so many contemporary products short-lived and un-sustainable.

The Heart of Sustainability: It is an object that relates to a broadly acknowledged set of human understandings that are independent of culture, religion, language or era; what Leibniz called the *philosophia perennis* (45), and Lewis referred to as the *evangelium eternum* (46). This undoubtedly contributes to its enduring and widespread use. However, one could say the same thing about a ceramic pot. So what is it that distinguishes one enduring artefact from another and makes it such an intensely personal and precious possession?

Objects that have a wide range of characteristics and meanings, including the profound, greatly surpass those of basic, utilitarian goods and this is what makes prayer-beads, and not pottery, so important for our understanding of sustainability. It is an artefact that has been conceived in response to our highest needs, which have been termed "self-actualization" and "transcendence" (47) – and which refer, respectively, to attaining one's potential and relating to something beyond the ego (48). In addition to these higher intentions, prayer-beads also reference other needs, such as social standing and identity (49). They also serve a basic function and have a variety of meanings related to popular culture, e.g. talisman. Thus, they can be understood, used and acknowledged in many different ways.

An Essentially Personal Object: The intimate "personal-ness" of the ownership of this object is a rare but very important characteristic to bear in mind when considering the nature of sustainable objects. When we value a particular object in a deeply emotional and personal way, then it becomes precious to us and worthy of our care.

A Challenge to 'Localization' and its link to Sustainability?: There has been much discussion about the need for increased 'localization' to contribute to sustainability in product design and manufacturing (50,51). However, to an extent at least, prayer-beads would appear to challenge this claim. It is certainly true that in many parts of the world this object is made at the local level of local materials such as plant seeds. However, it is also mass-produced from inexpensive 'anonymous', un-symbolic materials, and yet can still hold a profound meaning and a deeply intimate sense of 'possession-ness' for its owner. This is because the locus of this sense of ownership is related more to what the object represents, or to that which it points, rather than to what it actually *is* in terms of its materials or mode of manufacture. Any detrimental reaction due to its cheap, ubiquitous 'thing-ness' is overcome by its iconic associations, so that it can still be a deeply meaningful and intimate personal possession. This is perhaps the most important lesson for sustainability. *The **meaning** of an object, even of a newly manufactured, mass-produced plastic object, can provide a deep sense of ownership and value and can eclipse the specific physical characteristics and any physical shortcomings of the object.*

From this it seems reasonable to draw a further conclusion. At its most basic, utilitarian, 'un-designed' level, we could say that a functional object is capable of fulfilling an identified human need. Once we go beyond this basic utility and introduce 'design', to give the product market-appeal, then we start assigning to the product facets that will, ostensibly, satisfy a range of other human needs, such as 'a sense of belonging', and 'self esteem' needs. Objects designed to appeal to these needs, i.e. *'functional, social/positional* goods' are often rapidly outdated and unsustainable. Beyond these 'middle-level' needs, however, there are the higher needs such as aesthetic and spiritual needs. Products conceived to refer to these can appeal to our highest potential and in doing so *the very factors that spur unsustainable practices in objects are overcome.* In the one example of prayer-beads at least, we have a product that is inherently sustainable, more than simply functional, and ubiquitous. This example demonstrates that this combination is at least possible to achieve. The challenge is to see if it is possible in more common, everyday products.

At this point, we make try to take a few steps beyond the example of prayer-beads, to include some less explicitly religious products that are, at least to some extent, simultaneously *functional, social/positional,* and *inspirational/spiritual.* It is difficult to give such examples, and any selections will inevitable be subjective and perhaps contentious. However, they might include some of the work by Philippe Starck, such as his *Juicy Salif* lemon squeezer of 1990 for Alessi. This product may not be especially functional, and its prime role would appear to have become positional, but it is also a strikingly sculptural and perhaps inspirational design. Similarly, the designs of Daniel Weil, Ron Arad and the Droog designers are not merely functional, nor are they simply a combination of function and social/positional characteristics. The sculptural and aesthetic attributes of these designs tend to endow them *inspirational/spiritual* qualities.

These examples are perhaps not ideal, their durability has yet to be tested, and, in some cases, it is often difficult to rise above their strong *positional* associations. However, they do provide some indication of direction. These examples combine the various product characteristics discussed above, and encapsulate meanings, beauty and sculptural qualities that allow them to rise above the mundane.

In an attempt to combine *function* with *social/positional* and *inspirational/spiritual* characteristics in a broadly even way, and in a manner that addresses the more generally recognized characteristics of sustainability – such as enduring design, use of recycled materials, and integrated scales of manufacture (from mass-production to local scale), the author has developed a number of experimental designs. These, in effect, reverse the usual role of the designer. The product enclosure is replaced by a supportive rectangle of white canvas, and the functional components are composed on this rectangle, see figs. 5 and 6. In this way an attempt has been made to eschew the usual 'styling' of products, which is so closely associated with fashion and product transience.

7 CONCLUSIONS

Many of our contemporary products go beyond basic utility, to include a multitude of technical features, and styling and aesthetic considerations. The vast majority of these products are short-lived, un-repairable and, by any measure, un-sustainable. Given this state of affairs, we are faced with the question, "Is it possible to have an object that is more than merely functional, but which can also be understood as sustainable, and if so, what would be the characteristics of such an object?"

This discussion has attempted to answer this question and has revealed that sustainable product design is not to be found simply in the physical definition of an object, in the types or scales of manufacturing, or even in the nuances of the design. It also suggests that sustainability does not necessarily *require* a return to local production, the use of natural materials or high-value materials, craft-processes, or even high-quality production. Instead, once basic utility is surpassed, we enter an area of design that deals with the social and positional aspects of material culture and it is this area, when added to function, that appears to stimulate consumerism, disposal products and unsustainable practices. Furthermore, beyond the *social/positional* lies another area of human understanding, *the inspirational/spiritual*, that seeks and brings higher meaning to our endeavours. When this level of understanding informs our material productions, the destructive tendencies within the *social/positional* can be overcome and lead to objects that are, in their fundamental conception, deeply meaningful. And it is only by attempting to make our material culture meaningful that we can hope to contribute to a sustainable future.

Table 1. Product Characteristics

PRODUCT	
CHARACTERISTIC	
Inspirational/ Spiritual	• Symbolism, Representation • Allusion, Reference • Sacred, Religious
Social/ Positional	• Symbolism, Status • Social Standing • Decoration, Appearance • Taste. Trends
Functional	• Utility, Function • Usefulness • Human/Object Interaction • Object

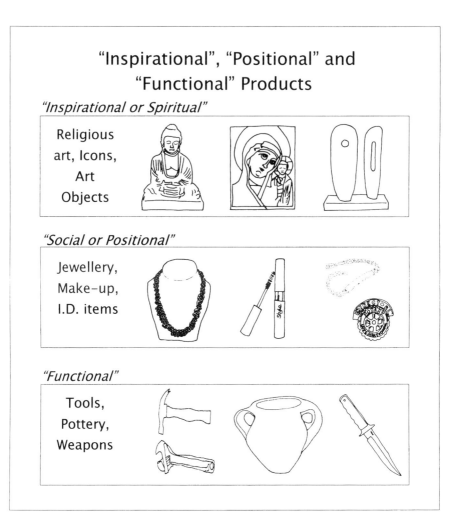

Figure 1. Inspirational, Positional, and Functional Products

"Inspirational + Positional" and "Functional + Positional" Products

"Inspirational + Positional"

Ornaments, Souvenirs, and Decorative objects related to social

"Functional + Positional"

Cars, Electronics, Hi-tech objects,

Consumer goods

Figure 2. "Inspirational + Positional" and "Positional + Functional" Products

"Inspirational + Positional + Functional" Products

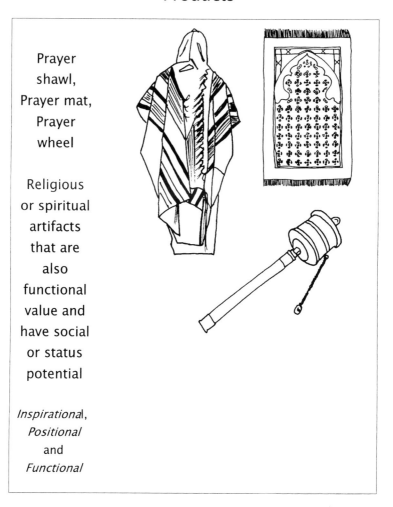

Prayer shawl, Prayer mat, Prayer wheel

Religious or spiritual artifacts that are also functional value and have social or status potential

Inspirational, Positional and *Functional*

Figure 3. "Inspirational + Positional + Functional" Products

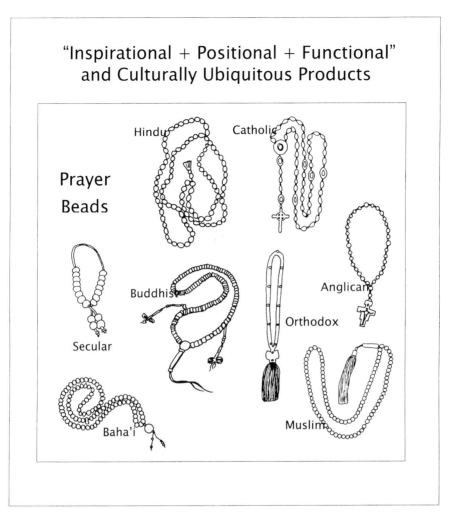

"Inspirational + Positional + Functional" and Culturally Ubiquitous Products

Prayer Beads

Hindu

Catholic

Anglican

Orthodox

Buddhist

Secular

Muslim

Baha'i

Figure 4. "Inspirational + Positional + Functional" and Culturally Ubiquitous Products

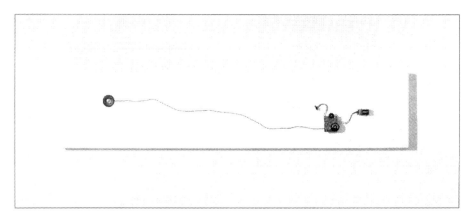

Figure 5. White Canvas AM Radio

Figure 6. White Canvas Digital Clock with Fruit Battery

D004/017/2004

REFERENCES

1 **Sachs, W., Loske, R., Linz, M.,** *et al* (1998) Greening the North: A Post-Industrial Blueprint for Ecology and Equity. Zed Books Ltd., London, p.110.

2 **Hawken, P.** (1993) The Ecology of Commerce: A Declaration of Sustainability. HarperCollins, New York, p.148.

3 **van Hinte, E.** (ed.) (1997) Eternally Yours: Visions on Product Endurance. 010 publishers, Rotterdam.

4 **Manzini, E. and Jégou, F.** (2003) Sustainable Everyday – Scenarios of Urban Life, Edizioni Ambiente, Milan.

5 A number of national museums have online collections – such as the British Museum at http://www.thebritishmuseum.ac.uk/, and the National Archaeological Museum of Athens at http://www.culture.gr/

6 **Amos, J.** (2004) Cave Yield 'earliest jewellery'. BBC News Online, accessed at http://news.bbc.co.uk/2/hi/science/nature/3629559.stm, 4/15/2004 7:27pm.

7 **Betts, K.** (2004?) Positional Goods and Economics. Lecture Notes. Swinburne University of Technology. Australia, accessed at: http://home.vicnet.net.au/~aespop/positionalgoods.htm, 4/14/2004 2:59pm.

8 **Lansley, S.** (1994) After the Gold Rush – The Trouble with Affluence: 'Consumer Capitalism' and the Way Forward. Century, London, pp.98&103.

9 **Postrel, V.** (2004) The Marginal Appeal of Aesthetics – Why Buy What You Don't Need. Innovation, The Journal of the Industrial Designers Society of America, Dulles, Virginia, Spring 2004, pp. 30-36.

10 **Hick, J.** (1989) An Interpretation of Religion – Human Responses to the Transcendent. Yale University Press, New Haven, pp. 129-171.

11 **Walker, S.** (2001) Games on a Stone Pavement: Design, Sustainability and Meaning. A lecture presented at The Royal Society of Arts, London, October 2001. Available at: www.rsa.org.uk/events/details.asp?EventID=1043

12 **Postrel**, p.36.

13 Objects that have *functional* and *spiritual/inspirational* characteristics, without possessing some *social/positional* qualities are probably impossible to find. This conclusion would correspond to Maslow's suggestion that human needs are hierarchical, in which case objects that have both *functional* and *inspirational/spiritual* characteristics, would also possess some *social/positional* qualities.

14 **Howard, K. L., and Pardue, D. F.** (1996) Inventing the Southwest: The Fred Harvey Company and Native American Art. Northland Publishing, Flagstaff, Arizona. p.7.

15 **Papanek, V.** (1995) The Green Imperative – Natural Design for the Real World. Thames and Hudson, New York, p.234.

16 **Numbers 15:39**, (1979) The Holy Bible – New International Version. Hodder and Stoughton, London, p.174.

17 **MacKinnon, M.** (2003) Would-be Warriors Return from Abroad – Iraqi Call to Arms. Globe and Mail, Toronto, A3, Wednesday, April 2nd, 2003.

18 **Apostolic Letter** (2002) Rosarium Virginis Mariae of the Supreme Pontiff John Paul II to the Bishops, Clergy and Faithful on the Most Holy Rosary, Section heading: October 2002 - October 2003, dated: 16th October 2002, accessed at: http://www.vatican.va/holy_father/john_paul_ii/apost_letters/documents/hf_jp-ii_apl_20021016_rosarium-virginis-mariae_en.html#top, 19/10/2002 10:48am.

19 **Kennedy, M.** (2003) Artist Trims Tate Tree. The Guardian, Manchester, December 13th 2003, online edition accessed at: http://www.guardian.co.uk/print/0,3858,4818609-110427,00.html, 17/3/2004 3.08pm.

20 **Langland, W.** (14th century) Piers the Ploughman. Penguin Books Ltd., London (1966), 5.401 – 5.402, p.73. Middle English version quoted here available at: http://etext.lib.virginia.edu/cgibin/browse-mixed?id=LanPier&tag=public&images=images/mideng&data=/lv1/Archive/mideng-parsed, accessed at 25/05/2004 10:30pm.

21 **Gribble, R.** (1992) The History and Devotion of the Rosary. Our Sunday Visitor Publ. Div., Huntingdon, Indiana, pp.130 & 169.

22 *Ibid*, 169.

23 **Wilkins, E.** (1969) The Rose-Garden Game – The Symbolic Background to the European Prayer-Beads. Victor Gollancz Ltd., London, pp. 32 & 56.

24 **Gribble**, p. 166.

25 **Chidester, D.** (2000) Christianity: A Global History. HarperCollins, New York, p. 275.

26 **Bauman, L. C.** (2001) The Anglican Rosary, Praxis, Telephone, Texas, p. 4.

27 **Gribble**, p. 167.

28 **Ward, M.** (1945) The Splendor of the Rosary. Sheed and Ward, New York, pp. 7-9.

D004/017/2004

29 *Ibid*, p. 8.

30 **Pascal, B.** (1995) Pensées. Penguin Books, London, revised edition, Series II (The Wager), section 418, p. 125, including footnote.

31 **Wilkins**, p. 14.

32 **Kido, O.** (2000) Newsletter, August 2000. Rokuon-ji Zen Centre, California, article by Abbot Osho Kido, accessed at http://pages.prodigy.net/monkkido/news/august-2000.html, 2/10/2004 7:23pm.

33 **Sims, G.,R.** (1911) In an Alien Land. Jarrold & Sons, accessed at: http://www.thhol.freeserve.co.uk/simsalie.html, 5/25/2004 9:27pm.

34 **Herrigel, E.** (1953) Zen in the Art of Archery. Vintage Books, New York, p. 43.

35 **Needleman, J.** (1980) Lost Christianity. Bantam Books, New York, p. 212.

36 **Wilkins**, p. 29.

37 **Gribble**, pp. 131-132.

38 **Wilkins**, p. 29.

39 **Vail, A** (1995) The Story of the Rosary. HarperCollins, London, pp. 104-105.

40 **Perkins, B.** (2004) Bottom Line Conjures Up Realty's Fear of 13, accessed at http://www.realtytimes.com.rtcpages/20020913_13thfloor.htm, 01/03/2004 4:24pm. Example: Lufthansa, see Seatmaps accessed at http://cms.lufthansa.com/fly/de/en/inf/0,4976,0-0-780757,00.html 30/05/20041:25pm.

41 **Wilkins**, pp. 50 & 179.

42 *Ibid*, p. 49.

43 *Ibid*, pp. 26, 29-30, 48.

44 **Apostolic Letter**, p. 19.

45 **Huxley, A.** (1946) The Perennial Philosophy, Triad Grafton Books, London, (1985), p. 9.

46 **Lewis, C. S.** (1933) The Pilgrim's Regress, Fount Paperbacks, Collins, (1977), p. 171.

47 **Postrel**, p. 36.

48 **Huitt, W. G.** (2003) Maslow's Hierarchy of Needs, Valdosta State University, Educational Psychology Interactive, accessed at: http://chiron.valdosta.edu/whuitt/col/regsys/maslow/html 2/10/2004 7:03pm.

49 **Wilkins**, p. 50.

50 **Dresner, S.** (2002) The Principles of Sustainability, Earthscan Publications Ltd. London, pp. 161-164.

51 **Van der Ryn, S. and Cowan, S.** (1996) Ecological Design, Island Press, Washington, pp. 57 & 65.

Critical friendships – 'non-Government organizations' (NGOs) as a stimulus for sustainable innovation and change

C SHERWIN
Principal Sustainability Advisor, Business Program, The Forum for the Future, London, UK

ABSTRACT

This paper introduces the Forum for the Future and describes its approach and work chiefly conducted in the Business Program (FBP). As part of the UK's largest Sustainable Development charity, FBP works with companies using a 'partnership' model to create innovation and stimulate change. The paper begins by introducing the background and history of Forum, its missions and working practices. Specifically it will highlight the work we do in the role of a 'critical friend' to companies trying to grapple with the sustainable development agenda. In this role, we balance support, knowledge and capacity building with a more challenging stance, which strives for ongoing improvement. The paper discusses how we actively engage with organisations and business on real sustainability issues. It also explains some of the cutting edge research we do, introducing Digital Europe (the sustainability implications of ICT and digital technology), Limited Edition (sustainable marketing) and future work on Sustainable Innovation. In doing this, the paper hopes to illustrate how NGO's can be a stimulus AND partner for sustainable transformation, innovation and change by adopting a solutions-orientated approach - as well as indicating how the NGO agenda is shifted and developing over the years.

1 BACKGROUND

Founded in 1996 by environmentalists Jonathon Porritt, Sara Parkin and Paul Ekins, Forum for the Future is one of the UK's leading sustainable development (SD) charities. Our work is based on the conviction that many of the solutions needed to defuse the environmental crisis and build a more sustainable society are already to hand. As a result, it is well documented that NGO's and other stakeholders within the SD debate have now become more 'solutions-focussed' – a shift well illustrated in Forum's mission *'to accelerate the building of a sustainable way of life, taking a positive solutions-oriented approach'*. Our work with more than 150 companies, local authorities, regional bodies and universities aims to build their capacity to overcome the many barriers to more sustainable practice. We aim for nothing less than transformation-irreversible change.

Forum tackles issues as diverse as renewables and climate change, farming, finance, environmental accounting and the digital divide. Our charitable status (and SD goal) is built on a commitment to share lessons learnt with decision-makers and opinion formers. Our aim is to help deliver not only a healthy environment, but also a better quality of life, strong communities, and practical answers to poverty and disempowerment. Central to our business model and educational purpose across all our work is the idea of 'partnerships'.

1.1 Forum Business Programme (FBP)

Companies and industry at large are facing increasing pressures from legislation, civil society, customers and other stakeholders to respond to the sustainable development in meaningful ways.

There have been a number of quite public 'spats' between companies and NGO's over the many years of the SD debate - both in and out of court. Companies have always, and still do face increasing pressures to respond to the SD agenda and demonstrate continuous improvement, from both NGO's and now a wider audience of stakeholders and broader drivers. And ironically, many of those same companies are turning too, rather than away from NGO's for advice, input and partnerships.

Our Business Program develops corporate thinking and behaviour to make sustainable development an integral part of business practice. The way we do this is through trust-based partnerships with key players across the business sector. We work with company decision-makers, helping them to integrate sustainability into company strategy and build capacity for practical implementation. These one-to-one relationships help our corporate partners to improve their social, ethical and environmental performance. Partners are 'reviewed' via our corporate guidelines and we aim to restrict partnerships to companies we feel can help deliver a sustainable future or at least ensure continuous commitment to sustainability improvement.

1.1.1 How this work

As a complex issue, sustainable development requires an holistic approach and solutions. Our work crosses many disciplinary and subject boundaries, though the work with business is centred on a partnership model (mentioned earlier) and delivered through research, advocacy and advice & communication. This can be mapped and understood in the following way:

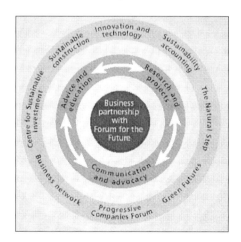

Fig. 1: The scope of work

- Partnerships - We work in one-to-one engagement with businesses through trust-based partnerships. We liase with key people in those companies, linking sustainability to company strategy and building capacity in sustainable development.
- Research - We develop cutting-edge research projects into sustainable development and business, promoting our research to decision-makers and opinion-formers in the business, research and policy communities.
- Advocacy - We aim to maximise the potential of our partnerships and research to influence government thinking at local, regional, national and international levels.

Each partner is able to draw on the experience and expertise of Forum's team of advisors through a work programme designed to address key areas of their business. Our partners also benefit from membership of a highly active Business Network, which meets at least three times a year. The Network encompasses all business sectors and provides a safe but challenging space in which to develop strategies and action plans. This is a dynamic forum that addresses diverse issues - from practical business tools for sustainable development to future challenges for business. All our work aims to educate the wider business community through sharing our learning – which is a fundamental purpose and requirement to maintain our charitable purpose. An example of this is Just Values – our report with BT, which explored arguments for looking beyond the business case for SD [2]. This discussion paper is not afraid to ask the most challenging questions on the business and sustainable development – that beyond the business case, this is actually a question of values, ethics and morality - arguments we are largely ignoring at present. This work is underpinned by crosscutting themes:

1.2 Critical Friendships
We try and engage with companies for a minimum of 3-years. But our role is not that of simply a passive deliverer of capacity (consultancy) or only as an external assessor of current practice (auditor) it is to challenge, stimulate and push organisations into new territories. Our partners have rather elegantly called this a 'Critical Friendship' – and we think this describes our aims, purpose and process rather well. We have found businesses tend to focus on the 'nuts and bolts' of business and can get caught up (or bogged down) in operational issues without considering the larger context and questions. For instance as we helped AWG develop a vision of what their organisation would truly look like in a sustainable future, which is providing a challenging roadmap with which to benchmark and assess their progress.

1.3 Transformation
Critical to our mission is the idea of real transformation towards sustainable development, and in fact this is a prerequisite of our partnerships. Our won internal research has demonstrated the growing number of NGO's now partnering with companies, though ironically a lack of transparency of and accountability for any results [6]. We annually measure the impact we have on partners via a Transformation Index that helps monitor how we are working towards our goal. This is a unique tool and process with which we can demonstrate the impacts and affects of our partnerships. The Transformation Index is conducted across 5 parameters, which are:
- How the relationship was managed?
- How we are connected to the partners core business?
- How capacity was built due to Forum?
- How the organisation has improved its SD performance?

- How the organisation has improved due to Forum?

The diagram below shows scores from our Transformation Index for the last 3 years for ALL Forum programs:

Total results across Forum ...

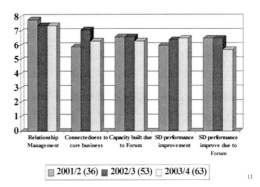

Fig. 2: The Forum Transformation Index

In general there are some small fluctuations in results over the years, but they remain stable. In certain cases you will see we have not achieved our own (rather challenging) target of a 7 ranking across all 5 areas, illustrating our own need to work on continuous improvement.

2 RESEARCH PROJECTS

Our research projects aim to develop cutting-edge thinking into sustainable development and business, promoting our research to decision-makers and opinion-formers in the business, research and policy communities. This usually tackles some of the pressing and current issues on the Sustainable Development radar. Recent projects are described below:

2.1 Digital Europe: ICT and Sustainable Development

"ICT, notably the internet and mobile telephone, are transforming the way we live and work, our social relationships and communities, even our impact on the environment. They are changing the very nature of human communication. (Digital Europe)... looks in detail at the defining characteristics of the digital society, and suggests that ICT equips us better than at any previous point in the industrial age to face the challenges of sustainable development"

Making the Net Work: Sustainable Development in a Digital Society
Alakeson et al, 2003 [1]

There are historically strong links between design and manufacture and new technology developments. In fact technology is often seen as a key (perhaps the key) drivers for product and process innovation. But how can we understand the sustainability implications of new technology developments? Such questions have been asked about digital technology and GMO's, and are beginning to do so about nano-technology.

Based on a consortium model of organisations which included - AOL Europe; EMI; Barclays; Sun Microsystems; HP; GeSI; Vodafone; Welsh Assembly Government, etc - Digital Europe was a European Commission (DG Information Society) funded project set up to investigate the relationship between Information and Communication Technology (ICT) and sustainable development, with particular emphasis on the role of and impacts on business (available at: http://www.digital-eu.org/). Research was conducted by three organisations, corresponding to the social (Forum for the Future), environmental (Wuppertal Institute) and economic (FEEM) dimensions of sustainable development. Primary and secondary research was conducted in close cooperation with the funding partners giving privileged access to information and key individuals.

If one believed the rhetoric from companies and promoters of the Digital Revolution over the past few years, ICT and digital technology should to be our environmental and social saviour, reducing impacts, dematerialising production, consumption and technology and effortlessly democratising socio-cultural infrastructures. But in reality, how is this technology revolution doing from a sustainability perspective. Is it moving us towards or away from sustainable development? The project aimed to answer this central question, and some results are discussed below:

2.1.1 Environmental Dimensions
There is as yet no evidence to suggest that the European economy as a whole is becoming more resource efficient as a result of the production and use of ICT. Nor are there any suggestions that this is likely in the near future. The ICT sector itself is not significantly more resource efficient than the rest of the economy, and so growth in the ICT sector is unlikely to have a positive effect on the resource efficiency of Europe. At a product level, research found that the miniaturisation of products is not a viable answer to the material intensity of economic activities, when taken alone, as smaller products performing the same function as larger products are actually relatively more resource intensive. And virtual products are likely to be much less resource intensive than their physical equivalents – though unsurprisingly these can be effected by consumer behaviour and use patterns.

2.1.2 Economic dimensions
Research was conducted to explore whether the use of ICT in businesses was causing a change in the geography of economic activity in Europe. Using econometric analysis, two conflicting trends were identified. Firstly, it was found that knowledge-based industries are tending to form clusters more readily as a result of the adoption of ICT. Secondly, it was found that other industries, where knowledge is not a key determinant for success, were tending to disperse more widely because of ICT. Thus, although there are signs that ICT is contributing to a redistribution of economic activity, the sectors that are most likely to underpin the future prosperity of Europe's knowledge economy are more likely than before to be concentrated in core zones.

2.1.3 Social Dimensions
Primary research with 100 businesses in six European countries showed that companies that were more advanced users of ICT were also more attuned to the idea of sustainable development. However, ICT was not contributing to better performance on social and

environmental indicators within these companies. Research identified eight areas where the use of ICT presented significant risks and opportunities for the business response to sustainable development:

- The increasing need for technological literacy
- Increased access to information
- Virtualisation of products and services
- Changing work patterns
- The creation of a distinct virtual space
- The network as an organising principle
- Increasing speed
- The increasing importance of trust.

2.2 Limited Edition: Unlocking the potential for sustainable marketing

"Today's central problem facing business is not a shortage of goods but a shortage of customers. Most of the world's industries can produce far more goods than the world's consumers can buy...
Because of (this) overcapacity, marketing has become more important than ever. Marketing is the customer manufacturing department"

Kotler 2003 [4]

Ignoring the argument that sustainable development (not marketing) is the answer to industrial overcapacity, unintentionally in his above quote, the godfather of marketing highlights the central dichotomy of marketing in sustainability-orientated world– that of limits. Marketeers have long been accused of promoting conspicuous consumption; 'massaging' needs and increased resource use for decades. However, a world without marketing is not realistic as it provides a valuable function in getting goods and services to people that match their needs – not to mentions it 'potential' to contribute to sustainability communication. So somehow these two seeming contradictory terms need to be reconciled. Our Limited Edition project aims to ask these challenging questions – doesn't marketing simply promote increased consumption and increased resource use - as well as challenging some of the perceived wisdoms of marketing practice – market research methods do not and cannot accurately determine intangible and poorly articulated secondary needs like environmental and social justice.

Our work with companies over the last eight years, has found that marketing departments have been among the slowest to engage with the Sustainable Development agenda and the last functional 'doors to open'. In organisations, often the marketing function holds the purse strings and has huge strategic input and influence. It is also often the gateway to customers and can massively influence what is communicated and perceived. From a sustainability perspective, we simply want more communication on the subject – and want business to be a key (arguably the key) driver for and educator on SD.

Limited Edition aims to unlock that potential Sustainable development offers radical new opportunities for customer insight and new product innovation, as well as reputation and brand enhancement. So far, we are or will be working with British Gas, Unilever and Vodafone in getting to grips with how to reconcile the satisfaction of consumer needs imaginatively – and profitably - with the needs of society, the planet and future generations. This takes place via a series of workshops placing sustainable development in a marketing context and help develop product and campaign implementation plans. This practical process of improving marketing practice and brands from the inside of companies is supported by an expert advisers group including Rita Clifton of Interbrand and ex-Saatchi executives Steve Hilton and Giles Gibbons, Professor Bob Worcester of MORI. Our partners keep in touch with each other through our Limited Edition network, and we will publish their collective experiences in 2005. Our experiences are showing:

- How to use sustainability in a strategic marketing context as a driver for new business idea generations.
- Ways in which SD can be connected to individual products and brands.
- The lack of maturity in market research methods – for capturing SD needs and consumer insights.
- Sustainable communication and how to identify the right messages and channels.

2.3 Sustainable Innovation

To achieve the proposed sustainability goals of Factor 4 or 10 reductions in resource consumption [7] and to drive radical and systems based 'eco-innovation' [3], new thinking and approaches are needed that go beyond our current reality. In this sense, perhaps more than anything else, sustainable development is a problem of and for innovation. Our mission, approach and impartiality make us an ideal partnering context to ask such fundamental questions, challenge and help conduct such work. Future research will focus on Innovation for Sustainable Development. At present we are considering going two ways via this work:

- Developing an integrated sustainability tool for product service assessment, evaluation and improvement – At present there are myriad (usually life cycle based) assessment tools for products and services. These tend to focus largely on environmental issues and impacts in product & service development. Sustainable development considers broader issues than only environment, so this project aims to develop a tool to understand the FULL sustainability impacts and implications of products and services.

- Where innovation does happen in business, it tends to be of an incremental, 'corrective' nature. This usually takes an existing product concept, understands its impacts (either a single one or the key impacts), and then improves these through product redesign (on recyclability, materials, energy use, etc). Whilst this has been successful it is unlikely to deliver sustainability. What is needed are totally new products and concepts and the thinking that goes with this. An example of this in non-SD areas is concept cars, one sees in car shows and exhibitions. Concept cars help capture the imagination or designers and audience without being limited by current technology or market expectations. In turn they help create demand and 'massage' desire. We aim to transfer this model of 'experimental' projects to sustainability - with the goal of a method for 'the concept cars of sustainability'.

2.4 NGO's and Innovation in Design and Manufacture for Sustainable Development

Inevitably our approach of being solutions-orientated, the emphasis of our working process on partnerships and our goal of creating transformational sustainability change will see and is seeing a greater emphasis on innovation – traditional territory for design and manufacturing. This is especially true of the partnerships with companies in the business program – as many partnering companies move beyond developing policies, strategies and programs, towards integrating sustainability into their core business processes, everyday activities and the 'nuts and bolts' of business where innovation usually sits. In this shift NGO's might play an increasing role in design and manufacture for sustainable development. This in turn will be (and is being) reflected in the types of work, partnerships and projects we undertake in the Forum Business Program – more of which is in collaboration with research and academic institutions. Critical friendships offer a useful company platform and cross-cutting theme for that, in that though we are unlikely to hold the pen, conduct the thinking or produce the idea, we can help.

3 CONCLUSIONS

This paper presents Forum for the Future, describing how it is creating sustainable transformation and working towards its mission 'to accelerate the building of a sustainable future'. Its shows how Forum and particularly its Business Program does this through partnerships actively engaging with and challenge participating companies. It draws particular attention to our research work by describing two current and one future research projects aimed to expand current knowledge and application of sustainability thinking. In all our work with companies our central role and stance (critical to our charitable status, our values and our external impartiality) is the role as a 'critical friend'. This has proved of great benefit to partners and a broader audience.

3.1 A new role for NGO's?

In many ways our work is a reflection of the new and emerging role of the NGO of the 21st Century – a more general trend within society and the economy mapped and identified recently [5]. This has seen a 'sea change' in the positioning of NGO's beyond only the traditional r ole a s campaigner and advocate – to having a much complex relationship with business and much more central and trusted role at the centre of 'civil society. In general within many NGO's, this greater emphasis on solutions and partnerships, and specifically to Forum, this role as a 'critical friend' is helping stimulate sustainable innovation and change in new ways and from new places unimaginable 5-10 years ago. Indeed solutions-orientated partnerships with business are likely to see a greater emphasis on innovation in product development, marketing and manufacturing in the next few years – another trend highlighted in the development of our research projects.

REFERENCES

1. **Alakeson, V., Alrich T., Goodman, J., and B. Jorgensen** (2003) Making the Net Work: Sustainable Development in a Digital Society. Xeris Publishing

2. **BT and Forum for the Future** (2003) Just Values: beyond the business case for sustainable development (available free from: http://www.forumforthefuture.org.uk/publications/JustValuesPublication_page760.aspx)

3. **Fussler, C. and P. James** (1996) Driving Eco Innovations: A breakthrough discipline for Innovation and Sustainability, Pitman Publishing, London.

4. **Kotler, P.** (2003) Marketing Insights from A to Z: 80 concepts every manager needs to know, John Wiley & Sons, Inc.

5. **SustainAbility** (2003) The 21st Century NGO: In the Market for Change. Sustainability, The Global Compact, United Nation Environment Program (UNEP) (available at: http://www.sustainability.com/publications/latest/21C-ngo.asp)

6. **Tuxworth, B., and F. Sommer** (2003) Fair Exchange? Measuring the impact of not-for-profit partnerships, internal publication (available from: http://www.forumforthefuture.org.uk/publications/Fairexchange_page1452.aspx)

7. **Von Weiszacker, Lovins, A. and H. S. Lovins** (1997) Factor Four: Doubling Wealth, Halving Resource Use. Earthscan, London

Strategies for
End of Life

Investigation into the use of engineering polymers as actuators to produce 'automatic disassembly' of electronic products

H HUSSEIN and **D HARRISON**
Brunel University, Egham, UK

Abstract

The need for an automatic and generic approach to the disassembly of electronic products to enable legislative targets for recycling to be met has prompted research into the development of 'self disassembling' products.

The use of shape memory materials to produce automatic disassembly of products to enable ease of recycling has been the focus of study of a number of projects at Brunel University [1,2]. However the prohibitive cost and unsuitable mechanical properties of these 'smart' materials has led to the need for research into new materials and methods.

A test program was constructed to evaluate the feasibility of using the shape recovery potential of standard engineering polymers to create actuator snap fasteners for mobile phones.

These snap fasteners provide a standard mechanism for holding/releasing product components during the products use phase. At end of life they can become actuators by the application of an appropriate trigger stimulus such as heat, allowing product disassembly to occur.

Dimensional constraints of mobile phone casings and the 'model' specific fastening solutions used in mobile phones provided an unrealistic design scenario for the development and implementation of 'automatic' snap fasteners. A generic mobile phone test bed was designed to allow flexibility and easy adaptation for a range of integral fastening solutions to be inverstigated. This allowed three design solutions for integral snap fastening systems to be trialled with minimum injection mould tooling costs.

The test bed design enabled different levels of complexity to be easily added to the prototypes enabling assessment of heat flow through the product with increasing levels of complexity. Successful disassembly of the prototypes was demonstrated in a pilot disassembly plant.

The results have shown that a number of standard engineering polymers can be successfully used to create automatically disassembling products. They have also allowed the current technological constraints of this approach to be defined.

1 INTRODUCTION

The shape memory effect in polymers is a function of the ability of the polymer to store strain energy that may subsequently be released after an elapsed time, when the appropriate pre-defined environmental conditions are met. In these studies only thermoplastics will be examined although this behaviour has also been well documented in certain thermosets.

The basis of the shape memory effect in polymers lies in the transition between the glassy and the amorphous phases. Polymers are classed as *viscoelastic* materials i.e. they exhibit material properties similar to both viscous liquids and elastic solids [3]. At low temperatures, polymers exhibit a glass-like behaviour, at higher temperatures they exhibit rubber-like behaviour and at even higher temperatures they behave like viscous liquids. At an intermediate temperature, the glass transition temperature (T_g), polymers are neither glassy nor rubber-like but exhibit a viscoelastic, intermediate modulus. It is this region that can be utilized to give us the shape memory effect in polymers.

If a polymer sample below its T_g is heated to T_g + (5 to 10)°C it may be deformed with little force. If this deformation is maintained and the sample is quenched to well below T_g this deformation will be 'frozen' in the sample. This 'embedded strain' is the basis for the shape recovery effect. Whilst in the glassy state the strain energy is insufficient to return the sample to its original unstressed form. However, on increasing the sample temperature to the region of T_g the rubbery state will allow the embedded strain to be released producing a shape recovery to the original form as long as the deformation does not exceed the elastic limit of the material or the sample is not constrained in any way that prevents this recovery. It is this behavior that allows the production of 'active' fasteners. These are fasteners that are in effect actuators and use the embedded strain energy to alter their shape and thus provide product disassembly.

2 HISTORICAL BACKGROUND

The production of Electronic and Electrical Equipment is one of the fastest growing world industries [4, 5]. The resultant Waste Electronic and Electrical Equipment (WEEE) stream has become an area of international concern. Electronic and Electrical Equipment Original Equipment Manufacturers are facing one of their most serious challenges with the introduction of European Union Legislation, such as the WEEE Directive 2002/96/EC, which comes into effect in August 2005[6]. This and other Extended Producer Responsibility legislation make products at End of Life the responsibility of the Original Equipment Manufacturers. Thus the challenge manufacturers face is how to apply an integrated approach to business and product development to create environmentally compatible products [5]. Within the European Union, WEEE has already reached 10-12 million tonnes a year with an increase of at least 3-5% expected per annum [4,6]. Due to their hazardous material composition, WEEE presents a number of environmental issues during the waste management phase if it is not properly treated. [4]. WEEE is diverse and complex in terms of materials, components, complexity and size [4,6] and consequently there are no generic treatment methods available. Mechanical/destructive recycling of Electronic and Electrical Equipment is the current state of the art technology but this is unsuitable to deal with the recycling quotas or the removal of hazardous substances. Disassembly is thus required in a majority of products before any other processing can occur. As manual disassembly is currently the only viable disassembly method there is a high cost associated with the end of life treatment of these products.

Automatic Product Disassembly is the technology of using shape memory alloys or shape memory polymers to create fasteners or fastening systems within a product. The application of a defined external stimulus will cause the fastener to change shape and allow the product to disassemble. In effect the 'smart' devices are actuators that cause disassembly by a change in the internal energy systems. The most common stimulus is an increase in the ambient heat above the trigger temperature. This ability to create self-disassembling products using smart materials provides a non-destructive, generic approach to product disassembly at end of life that requires minimal human input. In addition to shape memory alloys and shape memory polymers there are a number or standard engineering thermoplastics that exhibit degrees of shape memory effect.

Automatic product disassembly has been researched at Brunel University since 1996 as an enabling technology for end of life treatment of WEEE [1,2]. The principle of using the shape memory effect to produce active fasteners that will change shape or modulus on triggering to cause product self-disassembly, has been the subject of a number of consecutive research projects. Following a three and a half year EU Framework V project involving Motorola, Nokia and Sony, Active Fasteners Ltd has been set up to continue research into this field with a view to commercialization of this technology.

3 SHAPE RECOVERY EFFECT IN STANDARD ENGINEERING POLYMERS

3.1 Polymer Structure

The structure of a polymer has a significant effect its thermal behaviour. Polymers may be either crystalline i.e. the structure is three dimensionally ordered, or amorphous where the structure is unordered and contains no crystals. Completely crystalline polymers are rare. More commonly there are a number of amorphous regions (or domains) within the crystalline. These polymers are referred to as semi-crystalline polymers.

3.2 Thermal Transitions

Amorphous and crystalline domains respond differently to temperature variations. Crystalline and semi-crystalline polymers have a melting temperature, T_m, at which the lattice structure begins to disorientate without breaking of the chemical bonds. On cooling, crystalline and semi-crystalline polymers begin to form the thee dimensional crystal lattice structure at the

T_m. Amorphous polymers begin to stiffen at the glass transition temperature, T_g, as the long range thermal motion of the polymer chains stop.

3.3   Shape Recovery Effect of Polymers

For shape recovery to occur there is a need for a two-phase morphology to occur containing a 'Hard' domain (or block) and a 'Soft' domain. One of the domains (Hard) needs to be rigid and short whilst the other (Soft) needs to flexible and long. The hard blocks are the fixing phase and the soft blocks are the reversible phase. The hard phase dictates the level of rigidity, dimensional stability and thermal resistance. The soft phase, which can be either crystalline or amorphous, provides the elastomeric properties, primarily, recovery and energy absorption [7]. Heating the polymer allows softening of the hard phases that allows polymer flow. On cooling the hard blocks are re-established and the polymer is locked into this new shape. The soft elastomer phases provide the internal strain energy for recovery and on heating will recover to their original state as long as they are not constrained.

3.4   Shape Recovery Processing

A thermally stimulated shape recovery process may be defined as:

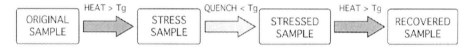

Figure 1. Thermal shape memory effect - Training and recovery process.

Where the original sample is heated to 5 to 10 °C above Tg and then deformed. The sample is held in the deformed position and quenched to at least 40°C below the T_g where the sample will now be 'frozen' in the new shape. This shape will be maintained until the sample is raised above T_g at which point recovery will occur.

From the pre-study of shape memory effect in engineering polymers [7] and some simple tests to ascertain the possibility of using engineering thermoplastics for active fasteners an empirical study was set up. This was to allow selection of a suitable material for active fastener solutions of integral snap fasteners in mobile telecommunication devices.

3.4   Empirical Study of Shape Recovery Effect of Polymers

3.4.1 Hypothesis

It was hypothesised that an ABS/Polycarbonate blended polymer [7] would exhibit notable shape memory effect properties in recovery of bending strain. From a set of initial tests, it was hypothesised that the degree of recovery would not be affected by the amount of bending strain applied a sample as long as the deformation did not exceed the elastic limit.

3.4.2 Empirical Investigation

An experiment was designed in order to test the hypothesis. Samples of two types of ABS/ Polycarbonate blends were prepared with sample dimensions of 10.0mm in width, 100mm in length, 3.5mm in thickness. A 'training' rig was fabricated which allowed the samples to be accurately bent to an angle of 90° at 5 different radii: 8mm, 10mm, 12mm, 14mm and 16mm. Test 1: The samples were placed in an oven at the glass transition temperature (T_g) for five minutes. The samples were then removed from the oven and immediately trained by bending at 90° around a brass rod of the required diameter, then immediately quenched in cold water.

A recovery test rig was designed and fabricated in order to allow observations of the angle of recovery of each sample. The sample was clamped in the recovery test rig and placed back into the oven, for 25 minutes, at T_g°C and allowed to recover. Readings on the degree of recovery were taken every minute.

It was noted that when training the samples, they were harder to bend than would normally be expected of a polymer that is at its glass transition temperature. Adaptations were made to this experimental technique in order to improve the validity of the results obtained. Therefore, a further set of tests was carried out (Test 2) where the sample was placed in an oven at T_g for 25 minutes in order to heat soak completely before training.

3.5   Results

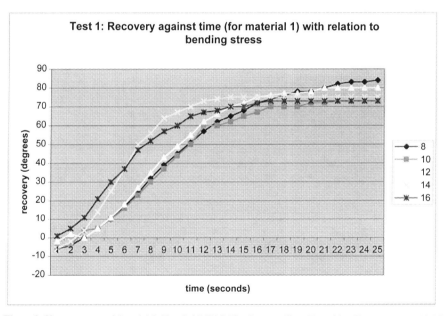

Figure 2. Shape recovery Material 1, Test 1:ABS/PC Blend   Glass Transition Temperature= 107°C

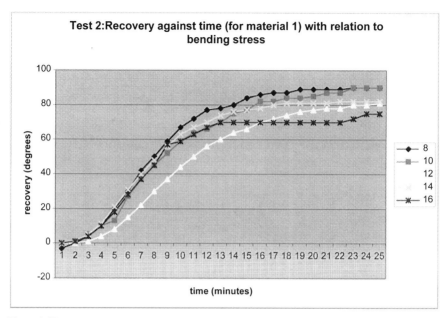

Figure 3. Shape recovery Material 1, Test 2: ABS/PC Blend   Glass Transition Temperature= 107°C

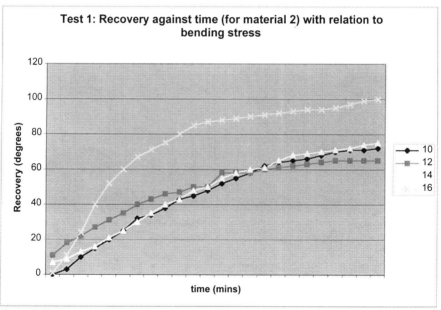

Figure 4. Shape recovery Material 2, Test 1: ABS/PC Blend   Glass Transition Temperature= 119°C

                                     D004/016/2004

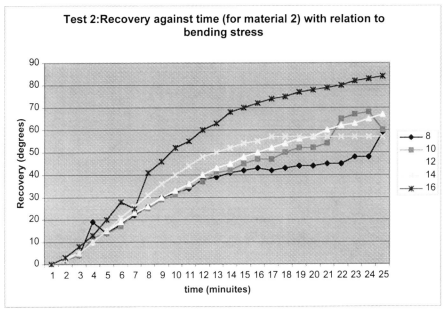

Figure 5. Shape recovery Material 2, Test 2: ABS/PC Blend   Glass Transition Temperature= 119°C

3.6 Discussion

For automatic product disassembly devices a useful shape memory effect is best demonstrated by the speed in which a polymer recovers. Selection of the most appropriate material was based on the degree of recovery between the samples first being placed in the oven up to the point where recovery decelerates, on average around 13 minutes. From the figures above it can be seen that Material 1 achieved a far higher rate and degree of shape recovery than material 2. This material was therefore chosen to fabricate the active fasteners. These results also indicated that for this material the maximum recovery that we could expect within 13 minutes was approximately 90% with the minimum recovery at the same elapsed time of only 65%. This dictated that any fastener designed should require no more than 90% recovery to enable disassembly.

4 ACTIVE SNAP FASTENER DESIGN

4.1 Design Limitations

Initially the aim was to implement designs for active fasteners into market products. However the complexity and tolerances to which mobile phone handsets are constructed makes integration of post manufactured disassembly solutions very difficult. Following a number of successful self-disassembling prototypes of basic solutions using shape memory alloys and shape memory polymers, the limitations inherent within this approach became apparent. An alternative approach to design, prototyping and testing of the disassembly solutions was required to assess the feasibility of using integral snap fit fastening systems in small hand held products.

4.2   Test Bed For Snap Fastener Solutions

Following discussions with Nokia Research Centre, Finland and the experience from working on early prototypes, it was decided that a generic box simulating a mobile handset was required that allowed a number of different solutions to be easily implemented. A simple mobile phone simulator model was developed that consisted of top and bottom covers and a chassis with recesses for a PCB and battery. The covers and chassis each had matching recesses to allow accurate and secure positioning of snap fastener components as shown below.

Figure 6. CAD models of bottom cover and chassis assembled plus top cover showing recesses for implementation of snap fastening solutions.

In order to increase the flexibility of the system the chassis was modelled as two mirrored halves that could be oriented either with the recesses facing each other so that the retaining lip would be on the uppermost and lowermost faces or with the recesses facing away from each other producing a centreline lip onto which snap fit clips could be oriented or attached. See Figure 7 below.

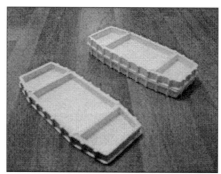

Figure 7. Chassis component and assembled version with centre lip

The design and manufacture of the injection moulded housing and the prototype snap fastening required a production run of no more than a few hundred parts and a high degree of adaptability within the manufacturing process. Whilst the parts needed to be injection moulded in order to replicate the materials and manufacturing processes, traditional tooling methods would have been prohibitive in time, cost and flexibility. The following method was therefore used to allow maximum control over these variables.

The parts were first modelled in a 3D CAD programme and fabricated using a fusion desposition method rapid prototyping machine. Prototypes were produced using an ABS base material with a claimed 70% material strength of injection moulded ABS and a model resolution of 0.1mm. The prototypes were then finished by hand and these models were then used as the patterns for resin casting the injection-moulding tool. A blank aluminium injection mould tool was prepared by completely removing the centre section to allow the pattern to be placed within the cavity and the resin be poured onto it. This method allowed two part moulds to be produced as below in Figure 8. or the simpler single part moulds for the snap fasteners. The resin used was aluminium filled epoxy resin with a maximum shrinkage of 0.1%, cast under vacuum and cured over 36 hours in a specific heating regime. The models were removed after the first low temperature curing. Following curing the moulds were polished and injection moulding carried out manually on a 22 tonne vertical axis injection-moulding machine.

Figure 8. Top cover, two part, aluminium filled resin injection mould tools

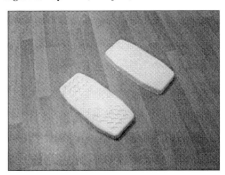

Figure 9. Prototype cover parts   Figure 10. 1ˢᵗ batch of assembled prototypes

4.3 Prototype Snap Fastener Production

The snap fasteners for the prototypes were produced in a similar way to the covers and chassis. However before an injection mould tool was made the solutions were tested for feasibility. Following design and rapid prototyping as previously outlined, silicon rubber moulds were made from the rapid prototype models and these parts were cast in a polyurethane two-part resin. These were attached to the casing and chassis allowing them to be tested for successful disassembly upon application of heat. Only following successful disassembly would the injection mould tool be produced. This approach allowed a quick assessment of solutions, with unsuccessful designs modified or rejected at an early stage. The polyurethane resin chosen gave a good simulation of the recovery properties of the ABS but without the mechanical strength or thermal properties. Three designs of clips were prototyped for final testing in a pilot disassembly facility.

4.3.1 Clip Design 1: Cantilever clip

One of the simplest forms of snap fastener is the cantilever clip. This was the first solution to be prototyped and tested in the test bed. The clips were placed at intervals around the top and bottom covers securing the chassis in position when on assembly by latching over the middle ridge (Figure 12). They were trained so that on application of heat, they straighten (Figure 14), moving away from the ridge with all three parts being released. The Figures below (11,12,13) illustrate the assembly and implementation of this solution.

Figure 11.   Arrangement of cantilever clips

Figure 12. Assembly of chassis and cover

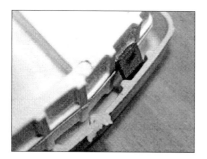

Figure 13. Detail of cantilever clip implemented

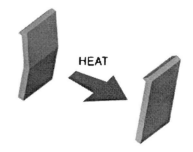

Figure 14. CAD model of shape recovery

4.3.2 Clip Design 2: Triple clip

A more complex version of this cantilever clip was developed. This solution was made up of three different clips though only one part was programmed. Figure 15 shows how the components interact to fasten the two covers to the chassis. Part 1 fixes on to the ridge around the inside of the chassis. Part 2, fixes onto the bottom cover and half of the length of the retaining edge snaps onto Part1, holding the chassis and bottom cover together. Part 3, which is attached to the top cover clips onto the remaining half of the retaining edge of Part 2. Part 2, is programmed part. As in the simple cantilever clip above, on application of heat it straightens releasing the other clips and thus all of the prototype components.

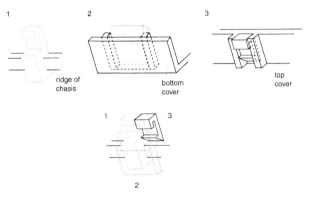

Figure 15.   Assembly diagram of Triple clip parts

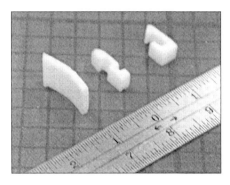

Figure 16.   Prototype Triple clip parts   **Figure 17.   Assembly of two Triple clip parts**

4.3.3 Clip Design 2: Wing Clip

The third solution prototyped concerned changing the orientation of the recovery movement with respect to the test bed components. A clip with two active parts was produced where the retaining face of the fastener and the recovery movement was in line with the longitudinal axis of the product rather than in the vertical axis, as with the cantilever type solutions. The figures below show the disassembly mechanism and implementation of the solution.

Figure 18. Wing clip prototype

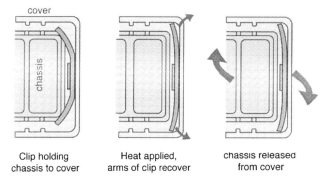

| Clip holding chassis to cover | Heat applied, arms of clip recover | chassis released from cover |

Figure 19. Shape recovery and disassembly mechanism

4.4 Active Disassembly Pilot Plant

In order to test the solutions as part of the Framework V project an automatic disassembly pilot plant was built by Gaiker Technological Centre and Indumetal Recycling, research partners, based in Spain.

The hot air based system consists of a rotary drum inside a chamber with hot air supplied by an internal air heater. A fan circulates hot air inside the closed chamber and reduces the energy consumption, in addition a hot air recycling system was also installed.

The main technical features of the hot air system were:
- External dimensions: 1900 x 1650 x 2650 mm
- Internal rotary drum: diameter 700 mm and length 1000 mm
- Heating system: 24 kW oven, cross current flow with hot air recovery
- Temperature range: 50-200°C
- Rotating speed: 2-25 rpm
- Insulation thickness: 150 mm

In addition a hot liquid system was also set up with technical features:
- Chamber dimensions 500 x 620 x 200 mm

- Temperature range : 25-100°C (heating media - water)
- Temperature range : 25-175°C (heating media – water/glycerin)
- Power: 2.0 kW

5 RESULTS FROM THE AUTOMATIC DISASSEMBLY PILOT PLANT

Ten prototypes of each solution were produced and tested at the facilities of Gaiker and Indumetal. The prototypes were not damaged during the disassembly process although some distortion of the polymer housing was evident at the highest temperatures and there was no mechanical interference to aid disassembly.

Disassembly Solution	Hot Air Activation Time @ Temp°C	Hot Liquid Activation Time @ Temp°C
Cantilever Clip	~10 min @ 140°C	~1 min @ 117°C
		~14 s @ 165°C
Triple Clip	~ 6 min @ 140°C	~10 s @ 165°C
Wing Clip	No disassembly @ 140°C	~5 min @ 120°C
		~34 s @ 172°C

Table 1. Results from Active Disassembly Pilot Plant

Although disassembly did not occur in the air activated wing clip models the wing clips did recover. Disassembly was hindered due to deformation of the covers and the internal tolerances not allowing full recovery.

6 CONCLUSIONS

The aim of this research was to assess the feasibility of using engineering polymers to produce active fasteners to allow automatic disassembly of a product. From the results it can be seen that the shape recovery effect in the selected ABS/PC blend may be used successfully to achieve product disassembly. Due to the low rate of heat transference in the plastic housing restricting the heat to the active components product design has a significant part to play in the method chosen for automatic product disassembly. Hot liquid was found to be the most effective solution for the hand held size prototypes that were examined. Whilst the use of a test bed for assessing the validity of solutions was successful, further research using data collected from this and other tests will focus on modelling the shape memory effect in polymers. This will allow the production of predictive software to further reduce cost and time in design and implementation of new solutions.

7 FURTHER WORK

7.1 Long Term Stability

At present little is known about the long-term stability of the shape recovery effect in engineering polymers. In viscoelastic materials where a sample is subjected to a constant strain a decay in the stress is observed. The limiting factor in the stress relaxation will be the level of viscous flow. Where viscous flow occurs the stress can decay to zero over sufficiently long periods of time. Where there is no viscous flow the stress decays to a finite value. This decay in stress has important implications for the degree of recoverability of a trained sample and thus the disassembly of a product over a long time span. Test samples have been trained and stored, with a view to ongoing research being carried out in this area.

7.2 Creep compliance

It is also possible that over extended periods, snap fasteners under high loading at elevated temperatures will creep. If this occurs the resultant deformation may cause the shape recovery effect to be distorted in such a way as to affect the disassembly process. Further studies are needed into modelling the effect of creep on the shape recovery effect in polymers, also product specific automatic disassembly solutions will need to be examined.

8.0 ACKNOWLEDGEMENTS

To the European Union for funding the Framework V project 'Active Disassembly Using Smart Materials' and the project partners, GAIKER Technological Centre, Indumetal Recycling, S.A., Motorola, Nokia, Sony International (Europe) GmbH and the University of Stuttgart (IKP) without whom this work would not have been possible.

9.0 REFERENCES

1 **Hussein, H.,** (2002) Introduction to ADSM. Going Green, Care Innovation 2002 - Invited Session

2 **Chiodo, J.D., Billett, E.H. and Harrison, D.J.** (1999) Active Disassembly Using Shape Memory Polymers For The Mobile Phone Industry. IEEE

3 **Ward, I.M.** (1983) The mechanical properties of Polymers. 2nd Edition. ISBN 0-471-90011-7

4 **Cui, J., Forssberg, E.** (2003) Mechanical recycling of waste electronic and electrical equipment: a review. Journal of Hazardous Materials B99, pp. 243-263

5 **Kopacek, B., Kopacek, P.** (1999) Intelligent Disassembly of Electronic Equipment. Annual Reviews in Control 23, pp. 165-170

6 **Irasarri, L.M., Malaina, M., Fernandez, F.** (1995) Disassembly and Recycling of Electronic Consumer Products: An Overview. Technovation, 15(6) 363-374

7 **Ristolainen, N.** (2002) Prestudy of Shape Memory Effect: Modification of ABS polymer. Confidential Report from Nokia Research Centre, submitted as part of ADSM Framework V project.

Remanufacturing – a key strategy for sustainable development

W L IJOMAH and **C McMAHON**
Department of Mechanical Engineering, University of Bath, UK
S CHILDE
Department of Computer, Communication, and Electronic Engineering, University of Plymouth, UK

ABSTRACT

Remanufacturing is a process of bringing used products to "like-new" functional state with warranty to match. It recovers a substantial proportion of the resource incorporated in a used product in its first manufacture, at low additional cost, thus reducing the price of the resulting product. The key remanufacturing problem is the ambiguity in its definition leading to paucity of knowledge and research in the process. Also, few remanufacturing tools and techniques have been developed to improve its efficiency and effectiveness. This paper addresses these issues by describing the findings of in-depth UK case studies, including, a robust remanufacturing definition and an analytic model of the generic remanufacturing business process for improving remanufacturing knowledge and expertise.

1 INTRODUCTION

The major cause of continued deterioration of the global environment is the unsustainable pattern of consumption and production, particularly in industrialised countries [1]. In Europe, a raft of legislation, of increasing severity, has been designed to reduce waste from a list of European priority waste streams. The concept of 'Producer Responsibility' requires original equipment manufacturers (OEMs) to "take back" an equivalent used product for each one sold. The significance of remanufacturing is that it combines profitability and sustainable development benefits by reducing landfilling, as well as the level of virgin material, energy and specialised labour used in production [2,3,4,5]. Research indicates that up to 85% of the weight of remanufactured products may be obtained from used components, and that such products have comparable quality to equivalent new products but require 50% to 80% less energy to produce [2]. Its economic benefits include having low entry barriers, and providing 20% to 80% cost savings in comparison to conventional manufacturing [6]. Companies will increasingly require remanufacturing expertise as it extends the life of used products and avoids costly landfilling. Because it profitably integrates waste back into the manufacturing cycle, remanufacturing offers producers a method of avoiding waste limitation penalties whilst maximising their profits.

The ambiguity in remanufacturing **definitions** is a major problem for researchers and practitioners. It causes extreme difficulties in undertaking effective research and in correctly disseminating knowledge about the process [7]. At the same time, many individuals are unable to differentiate between remanufacturing, repair and reconditioning and refuse to purchase remanufactured products because they are unsure of their quality. Remanufacturers also perceive the scarcity of effective remanufacturing-specific tools as a key threat to their industry [4] and research shows that there is a need for analytic models of remanufacturing [8]. This paper presents a robust remanufacturing definition and a comprehensive generic remanufacturing business process model that can be used to improve remanufacturing expertise.

2 THE RESEARCH METHODOLOGY

To ensure manageability of the research, its scope was limited to the mechanical and electromechanical sector of the UK remanufacturing industry. The definition of remanufacturing as *"The process of bringing a used product to like-new condition through replacing and rebuilding component parts"* [9] was adopted as a working definition which could be altered as the research uncovered further information. The eleven companies involved in the research were selected because their activities fitted the working definition.

The research was undertaken via a three-phase research approach that followed Eisenhardt's case study methodology [10]. There were three groups of case study companies, the Phase 1, the Phase 2 and the Phase 3 case study companies. Information about the companies is presented in Tables 1 and 2.

The Phase 1 research sought, firstly, to define remanufacturing and differentiate it from repair and reconditioning and, secondly, to describe remanufacturing so that others would correctly understand it. The first objective was achieved through literature survey and a series of one-day observational case studies where the researcher investigated the remanufacturing operation via observation of remanufacturing companies supported by interviews with key company personnel. Here, the working definition of remanufacturing was analysed by comparing actual remanufacturing, repair and reconditioning practices. The second objective was achieved by providing information to illustrate how the remanufacturing operation functions. This involved using observation of remanufacturing operations and interview of remanufacturing practitioners to obtain a list of company-specific flow charts of the remanufacturing operation. These flow charts were then compared so that similarities between them could be drawn out and used to develop generic flow charts.

The second research phase validated the information obtained in the Phase 1 research through four-week, in-depth participative case studies in new remanufacturing companies. The output of Phase 2 was clear illustration of the shortcomings of the working definition and an explanation of how the new definition augments it.

Table 1. The phase 1 case study companies

Company A Company A rebuilds rolling stock. Its capabilities range from remanufacturing (rebuilding to at least original specification from the customer's perspective) to reconditioning and repairing (rebuilding back to a range of satisfactory working condition that may be below the original specification).
Company B Company B rebuilds quarrying equipment. It provides a wide variety of engineering services including individual assignments, production runs of mechanical components and fabrications as well as a parts repair and remanufacturing service.
Company C A supplier of remanufactured products for the soft drinks and brewing industries with core activity in the supply of fully remanufactured process and packaging lines in the brewing and soft drinks industry.
Company D Company D is a transmissions remanufacturer and undertakes all three processes of remanufacturing, repairing and reconditioning.
Company E Company E remanufactures open and semi-hermetic compressors for the refrigeration industry.

Table 2. The phase 2 case study companies

Company F Company F remanufactures compressors for the refrigeration industry.
Company G Company G is an international supplier of new and remanufactured transmissions systems, electronic control units (ECU's) and replacement parts.
Company H Company H is company G's sister company and remanufactures large industrial transmissions very close to company G's UK headquarters.

The third research phase ensured that the research findings would be useful to others by presenting it in a format that both academics and remanufacturers could use to solve their remanufacturing-related problems. This was achieved by developing a robust model of the generic remanufacturing business process using the IDEF0 modelling technique. The rationale for developing a business process model was that business process modelling is known to be useful where "there is a need for a shared understanding of what the business does and also where information is required to assist improvement change programs" [11]. At the same time the model had to be generic because the research sought findings that would be valid to remanufacturers in general rather than to a specific one. IDEF0 was chosen as the most appropriate model technique because it provides a complete picture of a process in a concise and consistent manner [12] and has been used successfully in many areas of business process undertaking [13].

A key part of the model development process was the use of a Phase 3 case study to develop a company-specific model of the remanufacturing business process. One of the Phase 2 companies, Company F, was also the Phase 3 case study company. This was necessary to ensure that the initial model was developed in:

1. A genuine remanufacturing organisation. The Phase 2 Case study companies were known remanufacturers. In fact they were "A" class remanufacturers because they held remanufacturing contracts from Original Equipment Manufacturing companies (OEMs).
2. An environment where there was a high probability of finding characteristics that were common at least to all the case study companies. Because the Phase 2 companies had validated results from the Phase 1 companies it was likely that both sets of companies shared some characteristics.

Once a model that satisfied the Phase 3 company was obtained it was assessed against the practices of the Phases 1 and 2 case study companies to implement any alterations that would make it valid for a wider range of remanufacturers. The reason here was to enhance the model's scope of application towards being generic.

The last part of the research tested whether the research findings were valid and useful. This was achieved by exploring whether the research had obtained correct results that would be useful to practitioners. In this instance practitioners were remanufacturers and academics because they sought remanufacturing knowledge and expertise. This involved having a panel of practitioners, consisting of case study companies, non-case study companies and academics use the "validation by review" method [14] to assess whether the model satisfied the "The needs of practitioners" [15]. The validating criteria were the usefulness, sufficiency and clarity of the model.

3 MAJOR CASE STUDY FINDINGS

The major research findings were a robust definition of remanufacturing and a generic model of the remanufacturing business process. The definition was obtained from the Phase 1 research and was validated by the Phase 2 research. The validated definition was used as a stepping-stone for developing the model through the Phase 3 research.

3.1 A new robust definition of remanufacturing

3.1.1 Shortcomings of popular current definitions of remanufacturing
The inconsistency in the definition of secondary market processes and the ambiguity of remanufacturing definitions can be illustrated by examining two of the most popular definitions of remanufacturing, one by Amezquita et al. [16] and the other our working definition, by Haynesworth and Lyons [9]. Amezquita *et al.* [16] describe remanufacturing as *"The process of bringing a product to like-new condition through reusing, reconditioning, and replacing component parts".*

D004/024/2004

In the same paper they describe reconditioning as a process that is different from remanufacturing and that produces products that are inferior in quality to those produced by remanufacturing. However, since remanufacturers state that the quality of a product is governed by the quality of its individual components, a product that has within it reconditioned components can be described as remanufactured only if remanufacturing and reconditioning describe the same process. If, on the other hand, as proposed by Amezquita *et al.* [16], remanufacturing is indeed superior to reconditioning, then a product that has reconditioned components (i.e. components that are below the quality standards of remanufacturing), must itself be below the standards of the remanufacturing process. Such a product can therefore not be described as remanufactured. Because the definition above has not differentiated remanufacturing from reconditioning the authors believe that the definition by Amezquita *et al.* [16] is ambiguous. Our working definition of remanufacturing as *"The process of bringing a product to like-new condition through replacing and rebuilding component parts"* was published by Haynesworth and Lyons [9]. They go on to explain that *"Products that have been remanufactured have quality that is equal to and sometimes superior to that of the original product"*. The case studies undertaken during this research indicate that this bringing of remanufactured products to *at least OEM original specification* is one of the important factors that practitioners use to distinguish remanufacturing from repair and reconditioning. Because of this, it is proposed that Haynesworth and Lyons [9] have provided one of the most precise remanufacturing definitions. However, this definition does not provide a method for the purchaser to easily recognise that remanufactured products have higher quality than repaired and reconditioned alternatives, or that remanufactured products have similar quality to new alternatives. Because of this it is proposed that the Haynesworth and Lyons [9] definition of remanufacturing is also insufficient.

According to UK trade organisations, such as the Department of Trade and Industry (DTI) and Federation of Automotive Transmission Engineers (FATE), the legal performance requirement for secondary market products, where such regulations exist, stipulates guidance about minimum quality levels only and producers are held to account on the warranty that they give their products. The case studies showed that practitioners believe that a warranty serves as a guide to a product's quality. In fact, they stated that they give their remanufactured products at least the same warranty as the OEM equivalent as a method of indicating that the quality of their product is similar to that of the OEM equivalent. The practitioners believed that remanufacturing, repair and reconditioning involve dissimilar work content and produce products of dissimilar quality. They also believed that remanufacturing obtains the highest quality of products followed by reconditioning, then repair. The indicated that the operations could be differentiated using two factors:

1. The level of quality of the secondary market product when compared to that of an equivalent new product.
2. The standard of the warranty of the secondary market product in comparison to that given to the equivalent new product.

3.1.2 Proposed new definition of remanufacturing
The new remanufacturing definition augments that of Haynesworth and Lyons [9], by introducing the practioners' quality indicator of warranty thus allowing remanufacturing to be differentiated from repair and reconditioning on the basis of the quality of its products relative to that of the equivalent OEM product.

Table 3 presents the new definition along with the proposed definitions of repair and reconditioning. Table 4 shows the three operations on a hierarchy based on the work content that they typically require, the performance that should be obtained from them and the value of the warranty that they normally carry.

Table 3. Proposed definitions of the alternative secondary market processes

Remanufacturing
The process of returning a used product to at least OEM original performance specification from the customers' perspective and giving the resultant product a warranty that is at least equal to that of a newly manufactured equivalent.

Reconditioning
The process of returning a used product to a satisfactory working condition that may be inferior to the original specification. Generally, the resultant product has a warranty that is less than that of a newly manufactured equivalent. The warranty applies to all major wearing parts.

Repair
Repairing is simply the correction of specified faults in a product. When repaired products have warranties, they are less than those of newly manufactured equivalents. Also, the warranty may not cover the whole product but only the component that has been replaced.

Table 4. A proposed hierarchy of secondary market production processes

Remanufacturing
Work content: Greatest degree of work content because of the total dismantling of the product and the restoration and replacement of its components.
Performance: At least to OEM original performance specification from the customer's perspective.
Warranty: At least equal to that of new alternative.

Reconditioning
Work content: Less work content than remanufacturing, but more than that of repairing. All major components that have failed or that are on the point of failure will be rebuilt or replaced, even where the customer has not reported or noticed faults in those components.
Performance: Rebuilding of major components to a working condition that is generally expected to be inferior to that of the original model
Warranty: less than those of newly manufactured equivalents

Repair
Work content: Lowest work content as only specified fault need be attended to
Performance: Inferior to those of remanufactured and reconditioned alternatives
Warranty: Warranties of repaired products are less than those of newly manufactured equivalents and may apply only to the part that has been replaced or worked upon.

D004/024/2004

3.2 The generic model

3.2.1 The IDEF0 background and concept

IDEF0 is a process modelling technique that illustrates the component activities and flows of a system thereby helping the modeller to identify what activities are performed, how the activities are performed as well the rights and wrongs of the existing system. Its main advantage is that it enhances involvement and decision making using simplified graphical methods. IDEF0 was based on a well-established graphical language, the Structured Analysis and Design Technique (SADT), and was developed in the 1970s for modelling missile development activities for The United States Air Force. It was subsequently modified for business use and in 1993 was released as a standard for Function modelling in FIPS Publication 183 by the National Institute of Standards and Technology (NIST). Its benefits include helping in organising the analysis of a system, improving communication between the analyst and the customer and establishing the scope of an analysis. An example node of an IDEF0 model diagram [17,18,19], is shown in Figure 1.

The inputs (things transformed into outputs by the activity) are shown on the left side of the activity box. The input arrowhead points towards the activity box to indicate that the input data or object is going into the activity. An example of an input would be the material used in making a product.

The outputs (the transformed inputs) are shown on the right side of the activity box. The output arrowhead points away from the activity box to indicate that the flow is emerging from the activity. Examples of outputs include warranty and the product made by the process.

Controls are inputs such as constraints or rules that govern the conditions of the transformation, for example, technical skills, and legal requirements. These are indicated at the top of the activity box and their arrowhead points towards the activity box.

Mechanisms are the means by which the activity is performed, for example, robots, conveyors or people, and are illustrated below the activity box with their arrowhead pointing towards the activity box.

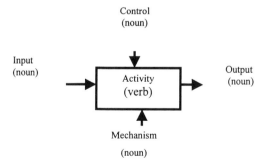

Figure 1. Example IDEF0 model diagram

3.2.1.1 Example of decomposition in IDEF0

The model shows a top-down decomposition from the context diagram. Figure 2 is an illustration of the use of decomposition to break an activity into its basic elements so that it can be examined in detail and fully understood.

The first level decomposition breaks the context diagram (A-0) down into subordinate activities. These subordinate activities may also be decomposed in the same way. There is no limit to the number of levels of decompositions. Each level of decomposition presents increasingly detailed information about the activity in question. The title of a decomposition diagram is taken from the box that it represents. Activities can be described as being parent or a child.

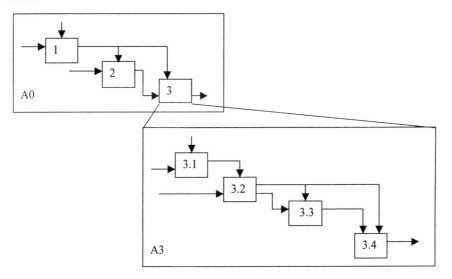

Figure 2. Decomposition in IDEF0

3.2.2 Description of the generic remanufacturing model

The model is a comprehensive document that unambiguously displays the resource required in all areas of the remanufacturing process, including the activities of all its sub processes, as well as the interrelationships between those sub processes. Its boundaries begin with the activities involved in the customer ordering a remanufactured product, goes through those involved in the company producing that remanufactured product, and ends with the activities of delivering the product to the customer. It consists of a series of embedded diagrams where top-level diagrams give basic overview of the system and lower level diagram give increasingly more detailed information. Because of this "Russian doll" characteristics it may be used as a tool for planning and controlling remanufacturing operations and could be used to help to design and implement effective and efficient remanufacturing businesses as well as to improve the efficiency and effectiveness of existing remanufacturing operations. For example, top-level diagrams give the macro-view of the remanufacturing process that top-level managers need to facilitate their strategic decision taking. The lower level diagrams provide detailed operational information to support shop floor workers in their everyday tasks. Figures 3 and 4 show the A0 and A-0 diagrams from the generic remanufacturing model.

Figure 3 is the basic diagram (A-0) of the environment of the remanufacturing business.

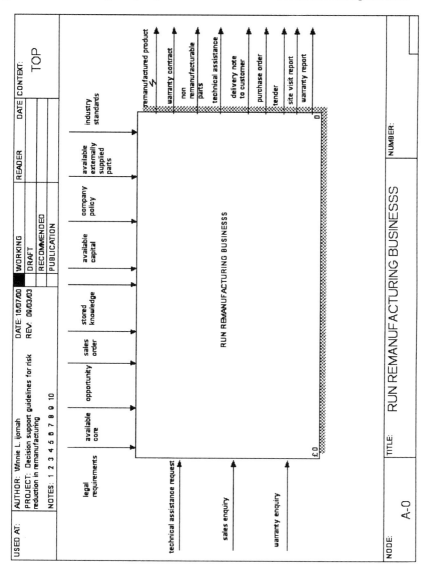

Figure 3. A-0 diagram

The A-0 diagram shows the interaction of the business with its environment. For example:

- Inputs such as technical assistance request, sales and warranty requests from customers
- Outputs such as remanufactured products and warranty
- Controls such as industry standards

This A-0 diagram can be decomposed to give A0 shown in Figure 4.

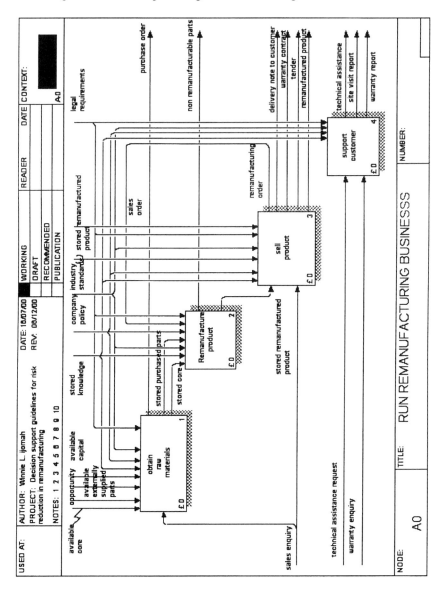

Figure 4. A0 diagram

The A0 diagram displays the four major remanufacturing activities: obtain raw materials; remanufacture product; sell product and support customer. Each of these major sub-activities are given with their various flows (inputs, outputs, control and mechanisms). They can also be decomposed themselves to reveal more detailed remanufacturing information.

4 VALIDATION OF THE GENERIC REMANUFACTURING MODEL

The model was validated by the review method [14] to assess its ability to satisfy the needs of practitioners [15]. If they found the model insufficient (a poor representation), unclear (incomprehensible) or inappropriate (unusable) then the research would have failed because the model would have been unable to fulfil the purpose for which it was developed. The validating panel was from the electromechanical sector of the UK remanufacturing industry and academics in remanufacturing-related disciplines because the research was geared towards them. All participants were middle management and above to ensure that they had adequate knowledge of the remanufacturing business process required for proper assessment of the model. The panel was independent of the research and the researcher's university and consisted of roughly equal numbers of academics, case-study companies and non-case study companies. This format permitted case study and non-case study practitioners to debate remanufacturing practices, and reach a consensus opinion in the event of anomalies being identified in the model by either type of remanufacturer. The validation was undertaken at the researcher's university so that the participants would not be distracted by their normal work duties. The close proximity permitted the panel's understanding of the IDEF0 modelling method to be monitored and also helped to ensure guiding of the discussion to ensure a systematic and rigorous validation. The researcher and research supervisor were present throughout the validation to take notes and to provide any additional support that participants may need by for example, answering queries and concerns. The information gathering media were white board, flip chart, tape recorder, common note taking and feedback sheets. There were two types of feedback sheets, the initial and the secondary feedback sheet. The panel handed in their initial feedback sheets, before leaving the session but retained the secondary feedback sheets which would be returned later with any further improvement suggestions that may emerge when they had discussed the model with their work colleagues. During the validation any amendment suggestions to model diagrams or to the model as a whole was recorded and debated by the panel to obtain a consensus opinion. The validation was successful because the model satisfied the validating criteria of usefulness, clarity and sufficiency. The validating panel believed that the model was accurate in its representation of the remanufacturing business process. They also found the model easy to understand and felt that it could help satisfy their requirements. All members of the panel believed that the model would be an effective tool for enhancing the efficiency and effectiveness of new and existing remanufacturing facilities.

5. Conclusion

The key remanufacturing problems are the ambiguity in its definition and the lack of remanufacturing tools and techniques, including analytic models of remanufacturing. This paper has addressed this by presenting a robust definition of remanufacturing and a comprehensive model of the generic remanufacturing business process. The definition for the first time differentiates remanufacturing from the related processes of repair and reconditioning to facilitate effective research and accurate results dissemination. The generic model displays all the resources and activities of remanufacturing in an unambiguous and easily comprehended manner and can be used as a tool for delivering remanufacturing knowledge and expertise as well as for analysing remanufacturing so that its efficiency and effectiveness can be enhanced.

The findings were successfully validated by practitioners using the review method [14], the validation criteria being usability, clarity and sufficiency as required by "The needs of practitioners" [15]. The validating panel indicated that the model was a good representation of remanufacturing and that they found it easy to understand and use and also that it could help them manage remanufacturing. The uses that the practitioners proposed for the model include training, documentation, and simulation. In the case of training, the model's advantage is that it does not rely on conversational language. This reduces ambiguity to allow people at whatever level of skills to receive the same quality of information thus improving communication and understanding of remanufacturing. In the case of documentation, as the model displays the interaction of the various company activities it could help employees to gain a whole system view of the company therefore promoting company-wide synergy. Such benefits would improve remanufacturing effectiveness and efficiency. The robustness of the findings was assured through the rigor of the research approach and the quality of the research design. For example, the research design stipulated extensive validation of all phases of the research by practitioners in consideration of the paucity of available remanufacturing publications against which the findings could be compared. Also, the finding passed the test for replication logic when assessed by practitioners and academics that were independent of the research and uninvolved in the actual case studies.

REFERENCES

1   **United Nations** (1999) "Sustainable development – Agenda 21" http://www.un.org/esa/susdev/agenda21.htm
2   **Lund, R.T.** (1984) "Remanufacturing: The experience of the U.S.A. and implications for the Developing Countries" World Bank Technical Paper No. 3.
3   **Lund, R.T.** (1996) "The remanufacturing industry: hidden giant" Boston University
4   **Guide** (1999) "Remanufacturing production planning and control: U.S industry best practice and research issues", Second International Working Paper on Re-use, Eindhoven, pp 115-128.
5   **Hormozi, A.** (1996) "Remanufacturing a nd i ts consumer, e conomic a nd e nvironmental benefits" APEX Remanufacturing Symposim, May 20-22, USA, pp. 5-7.
6   **McCaskey, D.** (1994) "Anatomy of adaptable manufacturing in the remanufacturing environment" APICS Remanufacturing Seminar Proceedings, USA., pp 42-45
7   **Melissen, F.W.** and **Ron, A. J. de** (1999) "Definitions in recovery practises" *International Journal of Environmentally Conscious Design and Manufacturing, 8*(2), 1-18.
8   **Guide** and **Gupta** (1999) "A queuing network model for a remanufacturing environment" Second International Working Paper on Re-use, Eindhoven, pp. 129-140.
9   **Haynsworth, H.C.** and **Lyons, R .T.** (1987) "Remanufacturing by design, the missing link" Production & Inventory Management; Second quarter, pp. 24 - 28
10 **Eisenhardt, K.** (1998) "Building theory from case study research" Academy of Management Review Vol. 14, No. 4: 532-550.
11 **Ould, M. A.** (1995) "Business processes: modeling and analysis for re-engineering and improvement" London; John Wiley and Sons [22] Zgorzelski and Zeno (1996) "Flowcharts, Data flows, SDT. IDEF and NIAM for enterprise engineering" The

international workshop on modeling techniques for business process re-engineering and benchmarking, 18-19 April 1996, Bordeaux, France., edited by Doumeingts G and Brown J. London: Chapman & Hall pp.71-81

12 **Smart et al.** (1995) "Guidelines for linking activity and information models of business processes-IDEF0 to IDEf1x" EPSRC Grant GR/J95010, working paper WP/GR/J95010-7.

13 **AMICE ESPRIT**, 1989, CIM-OSA Reference Architecture.

14 **Landry, M., Malouin, J.L.** and **Oral, M.** (1983) " Model validation in operations research"; European Journal of Operational Research 14, pp. 207-220

15 **Thomas, K.W.** and **Tymon, W.G.** (1982) "Necessary properties of relevant research: Lessons from recent criticism of organisational science" Academy of Management Review, Vol. 7 No3. 345-352

16 **Amezquita, T., Hammond, R., Salazar, M.** and **B ras, B.** (1996)," C haracterizing t he remanufacturability of engineering systems", Proceedings of ASME advances in design automation Conference, September 17-20, Boston, Massachusettes, USA, DE-vol.82, pp.271-278.

17 **Bennett et al.** (1995) " D ifferent types of m anufacturing processes and I DEF0 models describing standard business processes", Working paper WP/Gr/jJ5010-6, EPSRC Grant GR/J 95010, August 1995, Deliverable 3. UK

18 **Dorador J** and **Young R** (2000) " Application of IDEF0, IDEF3 and UML methodologies in the creation of information models" International Journal of Computer Integrated Manufacturing, 2000, vol. 13 No. 5, 430-445.

19 FIPS PUBS, 1993, "Integration definition for function modelling (IDEF0)", Federal information processing standards publication 183, National Institute of standards and technology, USA.

The automation of sustainability via re-use, modularization, and fuzzy logic

P BAGULEY, D BRAMALL, and **P MAROPOULOS**
School of Engineering, Durham University, UK
T PAGE
Department of Design and Technology, Loughborough University, UK

ABSTRACT

Modularisation of products makes re-use or update of some of its parts in the next product a greater possibility. Such re-use is preferable to material recycling when the corresponding energy consumption for recycling is a key environmental impact driver. Research has already been conducted into such an idea, and re-use is applied in cameras or copiers for example. Design engineers serve as an important source o f o pinion i n t he i dentification o f p otential modules. A fuzzy logic model consisting of a fuzzy measure, linguistic variables indicating the reliability of a data source, and a fuzzy integral is used.

1 INTRODUCTION

This research examines sustainable design in terms of environmental impact and cost. Sustainable design can take advantage of recycling or re-use of components, one sometimes being preferable to the other in terms of cost. In particular recycling sometimes means more cost than re-use because of the fact that more energy is needed than its alternative. It then becomes desirable to analyse scenarios in which re-use is made easier by adopting a suitable design approach. The approach used here is where re-use is made simpler through modular design. Modular design is the practice of designing products so that they can be broken down into independent functional or geometric units in order that units can be replaced by a pre-assigned policy. But how is this achieved? The solution used here is expert knowledge of re-use of an existing Component X as regards a target design or assembly, facilitated by information systems based knowledge. Fuzzy measures are a framework for capturing expert knowledge of "how much" an existing Component X can be re-used within a newly designed module or re-used compared to a collection of modules and sub-modules. Fuzzy sets can be used to capture expert knowledge about how relevant an expert's knowledge is in the light of certain data sources. A fuzzy integral can be used to synthesise expert opinion expressed in fuzzy sets and expert opinion expressed in fuzzy measures regarding the possible re-use of an existing Component X. The scheme is shown in Figure 1.

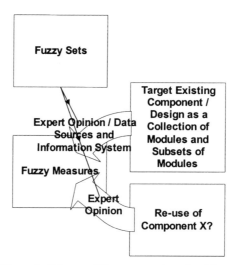

Figure 1. Scheme of the proposed methodology.

It is important to note that when capturing expert judgement in the fuzzy measure framework, and about how valued the expert opinion is itself, then a number between 0 and 1, or a fuzzy set describing a linguistic variable, might be used. In Table 2 a number is used for constructing the fuzzy measure and a linguistic variable is used for rating the expert judgement itself. This means the fuzzy integral used to synthesise the fuzzy measure and the rating requires fuzzy arithmetic.

2 SUSTAINABILITY AND COST

Much research within the subject of sustainability and cost has included life cycle cost and Life Cycle Assessment (LCA). This is not surprising since a sustainable design can be one, which impacts the environment in the least fashion but incurs high cost. Seo et. al. (1) provide a useful framework in which to consider life cycle costs (Table 1), although this framework does not provide a methodology of distinguishing between sustainability and life cycle cost. Indeed the problem is that it might be that the lowest life cycle cost is the least sustainable and creates terrible environmental damage that is not captured in the life cycle costs yet. New frameworks are being developed to better measure sustainability using cost, for example (2). It is therefore important for future research to determine the most relevant framework and framework elements for optimising sustainability through the measuring of life cycle costs.

D004/020/2004

Table 1. Life cycle cost factors (1)

	Company Cost	Users Cost	Society Cost
Design	Market Recognition, Development		
Production	Materials, Energy, Facilities, Wages, Salaries		Waste, Pollution, Health Damages
Usage	Transportation, Storage, Waste, Breakage, Warranty Service	Transportation, Storage, Energy, Materials, Maintenance	Packaging, waste, pollution, health damages
Disposal / Recycling		Disposal / recycling data	Waste, disposal, pollution, health damages

3 MODULAR DESIGN

The aim of modular design, in the context of sustainability and the environment, is to provide designers with a method of promoting re-use of components. Re-use is important in sustainable design because frequently recycling, a further strategy for sustainable design, expends more energy and hence cost, than re-use.

The objectives of modular design is a methodology for designers to identify modules to optimise the number of possibilities for re-use of an existing Component X in other and further designs, and hence spread the cost of components over a greater period of use (1), and promote less energy and material waste (2). The proposed methodology is one that automates the subjective reasoning process in design. In particular the methodology provides a framework in which designers can express their opinion about re-use and modularity in the design, but also have opinion expressed about how valued their opinion is in the first place.

Previous research has looked at how to promote modularity in design. Suh's axiomatic design (3) makes independence, a term frequently synonymous with modularity, a basic axiom. Suh's axiomatic design has been applied as a methodology for the modular design of a distributed cost engineering function (4). Here modularity is used to minimise the duplication and hence waste within the cost model development process tasks. Indeed modularity is important to be defined. Huang states: "modular products refer to products, assemblies, and components that fulfil various functions through the combination of distinct building blocks (modules)" (5). This work also discusses the activity of "externalising knowledge", i.e. making tacit knowledge implicit. This research shall utilise a case-by-case statement of what is the requirement for modularity, to motivate the use of expert judgement.

Kimura et. al. provide scenarios in which special cases of module identification can be highlighted, for example high turnover of technology around a module unit (6). They produced a solution by converting designs into a set of nodes and arcs denoting function and relations between nodes respectively. Overlaying several designs of existing components over each other is the idea that allows a frequency count of common functions and relations in designs. In this way, high frequencies point to modules within existing designs. This is indeed a possible methodology but does not go deep into the problem. It is possible that poor design intent is captured, in particular a frequently poor design. Also it is required that designs are stored and formatted correctly for using the method. An intelligent slant on the problem is lost through automation of a methodology that cannot escape its simple criteria of frequency and overlap of nodes. Hence this research seeks to best utilise the intelligence of designers, and not necessarily maximise their intelligence by matching it with an intelligent method.

This research aims to answer the question: "given an existing component, Component X, and a further whole assembly or newly designed assembly, how can the existing component be re-used in an existing or potential product?"

4 PROPOSED METHODOLOGY FOR MODULE IDENTIFICATION

The proposed methodology is to capture the intelligence of design engineers, supported by information systems, within the framework of fuzzy measures and fuzzy sets. The methodology is required to be integrated within the framework of a concept being used at the University of Durham, called Digital Enterprise Technology (DET). Examples of DET are e-business, Enterprise Resource Planning, and digital factory simulation. More formally DET is "the collection of systems and methods for the digital modelling of the global product development and realisation process, in the context of lifecycle management". DET is potentially introduced through product visualisation via a Computer-Aided Design package CATIA and its concept of design tables, and simple Excel spreadsheets.

5 WHAT IS FUZZY MEASURE THEORY?

The axioms referred to in this paper for fuzzy measures are a subset of all the axioms used in fuzzy measure theory. The other axioms are not mentioned. They refer to an infinite collection of sets or "ways of re-using component X" and are omitted because of space restriction. For a finite universal set, X and a family of subsets of X, C, then a fuzzy measure, μ is a function, $\mu :\rightarrow [0,1]$ following the axioms

1. $\mu(\text{empty set}) = 0$, and $\mu(X) = 1$
2. for all $A,B \in C$, if $A \subseteq B$ then $\mu(A) \leq \mu(B)$

The idea of a measure and a fuzzy measure is not straightforward to grasp. The problem faced by the engineer is to construct the fuzzy measure of how much a parameterised component can be considered for re-use by its fuzzy measure regarding a target design of assembly. For example how much is component X similar to a target design or assembly so that component X can be re-used instead of the target design or assembly. Function and geometry are examples of characteristics of how the fuzzy measure might be constructed. In light of the axioms above the empty set refers to no target design or assembly hence, trivially, component X cannot be re-used. The universal set X refers to an identical component X for the target design or assembly and hence, trivially, component X can be re-used in its current form as a direct substitution, hence the fuzzy measure is 1.

D004/020/2004

A theoretical example of a fuzzy measure is the Sugeno measure. The algorithm for the Sugeno measure is shown in Equation [1].

For subsets E and F of a Universal set X (for example ways E and F of using Component X):

$$g_\lambda(E \cup F) = g_\lambda(E) + g_\lambda(F) + \lambda \cdot g_\lambda(E) \cdot g_\lambda(F) \quad [1]$$

where:

E and F are formally crisp sets, g is the fuzzy measure and $\lambda > -1$. λ can formally be calculated from the axiom, $\mu(X) = 1$.

Research can find values of λ suitable for "component X for re-use" and associated target assemblies or design. It is intended that such values of λ can be found for categories of components and target designs and assemblies. For example an existing motor, in the category of motors can be related to the target design of a boat, this being in the category of boats. The fuzzy measure produced using the Sugeno measure shall provide a value of λ. It is hoped that λ can be plotted for different categories of components and target designs. The current research considers the above to be difficult and too formal.

6 PRACTICAL ISSUES

Having said that fuzzy measures satisfy a number of axioms as shown in Section 5, it is asked, but what does this really mean for an application to re-use and modular design? What we are dealing with in modular design is the contribution of data sources to the identification of modules for re-use. The objective of the research is to provide a framework and methodology for using data sources for such re-use and module identification (an example of a data source is an expert, or an expert utilising a further data source). It would be useful to have a "measure of uncertainty" in which the highest degree of certainty can be chosen. The situation with the most data sources and least uncertainty would provide the solution to the problem of module identification.

7 FUZZY SETS AND FUZZY MEASURES

Fuzzy logic made its inception in 1965 with the work of Lotfi Zadeh, of the Berkeley Institute of Soft Computing (BISC). The term fuzzy logic can be used to mean all manner of things. It is important in this work to draw a distinction between fuzzy set theory and fuzzy measure theory.

Fuzzy set theory is based on a set having a fuzzy boundary, rather than a crisp one. Therefore a set element belongs to the set as a matter of degree. In Figure 2, the number 4 belongs to the fuzzy set "about 6" to a degree 0.4. Fuzzy measures, the fuzzification of measure theory, studies how much a crisp set or element, not a fuzzy one, belongs to say a particular set from the power set of a collection of crisp sets. In Figure 3, the crisp (non-fuzzy) set B matches the crisp set B to a degree, say 0.6 (although the criteria for fit is arbitrary and not obvious). Reche and Salmeron give "importance" as a fuzzy measure, for example, where a finite set of "quality items" are measured for their "importance" (7). Here the empty set had no importance and the entire set (universal set) of quality items had an "importance" of 1, hence the first axiom is satisfied in Section 5. Fuzzy sets are the degree a single element belongs to a set, fuzzy measures are the degree a crisp set belongs to, or matches, one of the sets in the power set of a collection of crisp sets. Considering (7) then a degree of importance is attached to each member of the power set X, such that the degree of "importance" never decreases if an element is added to what is being measured, for example an addition of any quality item does not decrease the degree of importance of the existing quality items being considered for measure.

To further illuminate the use of fuzzy measures then a practical example follows. Let the set Y be such that Y = "the set of ways a Component X can be re-used regarding a target design or assembly". Such a set will be a reference set and might exist of the ways a, b, c, d and e. Therefore Y = {a, b, c, d, e}. A newly designed component might be able to be completely replaced by old Component X, or indeed a subset of the newly designed component replaced by Component X. A useful thing to do is, therefore, to take the power set of Y (all the ways Component X can be re-used), and let the designer make a judgement as to how much his newly designed component can be accommodated by Component X, or existing Component X can be re-used. The power set of Y has 16 elements ($2^{|Y|}$ or 2^4). The power set of Y is the collection of all possible sets formed from Y. Hence the power set of Y is useful for considering re-use of Component X through modularity as it considers the target product or design in terms of modularity. It is important to note that the first step of forming the modules of the target design or product is down to the engineer, and might well be potential or hypothesised modules.

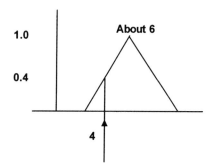

Figure 2. Demonstration of a fuzzy set.

D004/020/2004

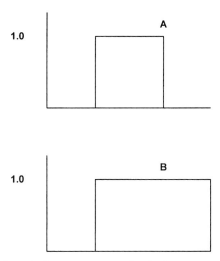

1.0   A

1.0   B

Figure 3. Demonstration of a fuzzy measure, what is the measure of B re-used as A?

8 PARTICULAR EXAMPLES OF FUZZY MEASURE THEORY

The axioms of fuzzy measure theory afford a framework in which to capture super-additivity or sub-additivity of the fuzzy measure. Super-additivity refers to the case where the measure of $A \cap B$ is greater than the measure of A plus the measure of B, sub-additivity is where the measure of $A \cap B$ is less than the measure of A plus the measure of B. In particular subjective opinion refers to a positive interaction of 2 subsets, or the inhibitive interaction of 2 subsets. The method, therefore, provides a useful framework and thinking tool, for capturing interaction, but is also only as good as the experts whom use it. For example, it might be that 2 different ways of re-using Component X as regards a target design or existing assembly, when considered at the same time, reduces the measure of re-use of Component X, and hence is sub-additive.

9 RESULTS

In the research a spreadsheet is used to record subjective opinion that is in turn used to build the fuzzy measure, as shown in Table 2. As can be seen the power set of the different ways Component X can be re-used are shown in the left hand column and the "evidence rating of expert" is rated using linguistic variables. The Choquet integral calculated for each row is found in the column labelled by E. It is assumed the rating of the expert is the same for all decisions he makes.

Table 2. Example spreadsheet of the proposed methodology.

Ways of Re-using Component X in a Target Design Module	Degree of fit of Component X for Re-use with Target Design (Expert 1)	Degree of fit of Component X for Re-use with Target Design (Expert 2)	Degree of fit of Component X for Re-use with Target Design (Expert 3)	Degree of fit of Component X for Re-use with Target Design (Expert 4)	Choquet Integral, E
a	0.56	0.78	0.23	0.4	?
b					
c					
a,b					
a,c					
b,c					
a,b,c					
Evidence Rating of Expert	Low	Low	Medium	High	

10 CHOQUET FUZZY INTEGRAL

A fuzzy integral is used to synthesise the fuzzy measure and the ratings of experts providing the measure. An example of a fuzzy integral is a Choquet integral. The Choquet integral uses the function representing the fuzzy measure to provide a new measure from the existing fuzzy measure of re-use and corresponding ratings of the expert whom are providing each value of the fuzzy measure. The general form of the Choquet integral is shown in Equation [2], where h corresponds to evidence supporting the fuzzy measure and g corresponds to the fuzzy measure itself (8). The Choquet integral is therefore used in connection with aggregating information (8). It is important to note that the algorithm is such that the $h(x_i)$ are arranged in decreasing order, thus that a new order is formed in the x_i (expert data source, i) corresponding exactly to the order of the $h(x_i)$. Equation [3] is the discrete version of [2] and shows how the output of the fuzzy integral ($E_g(h)$) is related to the change in evidence ($[h(x_i) - h(x_{i-1})]$) multiplied by the corresponding measure of suitability of re-use of Component X for the design or assembly under consideration (A_i), i.e., $g(A_i)$.

$$\int_X h(.) \circ g(.). \qquad [2]$$

where:
$h(x_0), h(x_1), h(x_2), .., h(x_n)$ is the evidence provided by $x_0, x_1, x_2, ..., x_n$. $x_0, x_1, x_2, ..., x_n$ refer to experts in this research.

$$E_g(h) = \sum_{i=1}^{n} [h(x_i) - h(x_{i-1})] g(A_i) \qquad [3]$$

D004/020/2004

The purpose of a fuzzy integral is to take the fuzzy measure and "fuse" it with the degree of confidence afforded in it by a range of data sources. These data sources may be experts directly or experts influenced by data sources. Research will find whether subsequently taking the highest value of the fuzzy integral gives the desired result of most suitable for re-use option, if the application is a ranking of alternatives. When re-arranging the order of the experts it is important to maintain the association with their opinion of the value of the fuzzy measure.

11 PART OF A WIDER SOLUTION

An Excel spreadsheet has been built that implements the proposed methodology and shall be used to rank potential solution modules for re-use by Component X. In particular it is hoped that the tool can be used to produce a "design table" for use with CATIA. A design table is an Excel spreadsheet that contains a number of configurations of a set of parameters. Each configuration is made by specifying a value each for a set list of parameters contained in the spreadsheet. Therefore different rows on a spreadsheet correspond to a variant of a parameterised design. Choice of parameters most conducive for product assembly re-use possibilities by Component X can be obtained from the re-use identification tool and fed into the design table to produce a design promoting component re-use.

12 FURTHER RESEARCH

Further research shall introduce re-use reasons, for example standardisation, commonality, function independence, cost, manufacturability, maintainability (6), as the evidence for re-use within the synthesis of the fuzzy integral.

13 SUMMARY OF THE PROPOSED METHODOLOGY

A summary of the proposed methodology is:

1. Identify the potential component, Component X, for re-use.
2. Split the target assembly or design into a set of "potential modules" to facilitate re-use of Component X.
3. Identify a team of experts to give their subjective judgement.
4. Form the power set of the set of "potential ways of re-use of Component X".
5. Capture the degree of match between the power set of "ways Component X can re-used" formed in step (4), and the target design or assembly modules, for each expert.
6. Rate the competence of the experts using fuzzy numbers between 1 and 10, where 10 is the best (these might be labelled with linguistic variables).
7. For e ach c omparison o r r ow i n t he E xcel table, order the ratings of competence of the expert / data source in decreasing order, and hence the data sources or experts used are ordered correspondingly. This is important for the Choquet integral so that the difference between ratings ($[h(x_i) - h(x_{i-1})]$) can be multiplied by the corresponding fuzzy measure ($g(A_i)$).

8. Use the Choquet integral to synthesise the fuzzy measure of re-use and the fuzzy sets used to rate the experts, for a particular set of potential modules and the component, Component X, for reuse.

9. The result is a value for each of the potential modules, and a ranking for which module or modules can be replaced by the potential component, Component X, for re-use. Further research will establish the efficacy of the ranking.

14 CONCLUSIONS

Fuzzy measures provide a framework in which to express design opinion. Choquet integrals provide a framework for the synthesis of opinion about data sources and also the opinion about which ways Component X can be re-used instead of newly manufacturing modules in a target assembly or design. The methodology is a proposed framework with which to rank how well a component can be re-used, for a module or modules of an existing target assembly or design. A sustainable design here is one which allows maximum re-use of existing parts like Component X.

REFERENCES

1. **Seo, K. K., Park, J-H., Jang, D-S.,** and **Wallace, D. (2002)** Approximate estimation of the product life cycle cost using artificial neural networks in conceptual design. *International Journal of Advanced Manufacturing Technology* 19, pp. 461-471.

2. **Howes, R.** (2002) "Environmental Cost Accounting: an Introduction and Practical Guide", Chartered Institute of Management Accountants.

3. **Suh, N. (2001)** "Axiomatic Design Advances and Applications", Oxford University Press.

4. **Baguley P., Wang, Q., Vogt, O., Qaqish, T., Bramall, D.** and **Maropoulos, P.** (2004) Systems thinking applied to life cycle costing and life cycle engineering. 10th International Conference on Concurrent Enterprising, 14-16th June, Seville Spain.

5. **Huang, C.** (2004) "A Multi-agent approach to collaborative design of modular products", *Concurrent Engineering: Research and Applications*, Vol 12 No 1, pp. 39-47.

6. **Kimura, F., Kato, S., Hata, T.** and **Masuda, T.** (2001) "Product Modularisation for Parts Re-use in Inverse Manufacturing", CIRP Annals Vol 50 / 1 / 2004, pp. 89-92.

7. **Reche, F.** and **Salmeron, A.** (1999) Towards an operational interpretation of fuzzy measures. 1st International Symposium on Imprecise Probabilities and Their Applications, Ghent, Belgium.

8. **Chiang, J.H.** (1999) "Choquet Fuzzy Integral-Based Hierarchical Networks for Decision Analysis", *IEEE Transactions on Fuzzy Systems*, Vol 7, No 1, February, pp.63-71.

Understanding Consumers to Create Sustainable Products

The 'recycled consumer' – evidence and design implications

P MICKLETHWAITE
Faculty of Art, Design, and Music, Kingston University, UK

ABSTRACT

This paper examines evidence of 'recycled consumption' and the 'recycled consumer' in the UK and elsewhere. Consumption of recycled products is the vital final link in the cyclical production and consumption system expressed by the idea of 'closing the loop'. 'Closing the loop' in the UK requires a significant increase on current consumption of recycled products. The contributions that design, in its various forms, can make are considered.

1 'CLOSING THE LOOP': THE CONSUMER

There is currently a great deal of political emphasis in the UK on recycling as a strategy for sustainability (1). UK household waste recycling collection rates must double between 2001/2002 and 2005 if the UK Government is to meet its aggressive recycling targets (1) (2). Collection of recyclable materials does not in itself constitute successful recycling, however. Production, marketing and consumption of recycled products is also necessary to properly 'close the loop' (3).

The UK Government, through the Waste and Resources Action Programme (WRAP), has been prominent in sponsoring the development of new high-volume commercial applications and markets for materials recovered from the waste stream. 'Closing the loop' requires that the products made from these recycled materials are then taken up by 'recycled consumers', for whom recycled products are a desirable and preferable option. Consumption of recycled products constitutes the final link in the cyclical production and consumption system expressed by the idea of 'closing the loop' (Figure 1). The recycled consumer therefore has a vital role in realising a closed loop cyclical production and consumption system. The availability of explicitly recycled products may also act as a spur to increased recycling collection, adding to the momentum of the recycling loop.

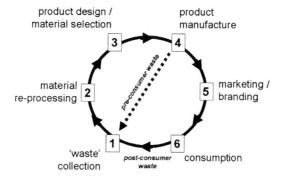

Figure 1. Cyclical production and consumption system

This paper continues the focus on consumption and the consumer introduced into the preceding DMSD conference (4). It complements the focus on the role of design in closing the loop in an accompanying paper submitted to the present conference (5).

2 THE 'GREEN CONSUMER'

The 'recycled consumer' can be seen as a subclass of the 'green consumer'. Green consumption introduces explicit environmental concerns into purchasing and consumption decisions.

> "'green' consumers ... consistently seek product or company information and attempt to integrate a variety of environmental and or societal influences with their buying behaviour." (6)

The emergence of the green consumer in the UK has been identified as a relatively recent phenomenon.

> "Opinion polls of the late 1980s evidenced the [UK] green consumer as an established entity." (7)

The first significant text on the green consumer published in the UK appeared at this time (8). A subsequent decline in self-proclaimed green consumption in UK polls conducted in the 1990s is identified, however, a decline attributed to growing consumer cynicism towards the accuracy and validity of companies' environmental claims (7). In addition to an apparent fall in poll respondents' self-reporting of their green consumption behaviour, an 'intention-action' gap has also been identified between proclaimed green consumption (the claims people make in surveys) and actual purchasing behaviour (what they actually buy) (9). Contemporary evidence from retailers apparently failed to support even the modest green purchasing claims made by consumers in the 1990s.

> "The experiences of supermarkets suggest that, despite opinion poll findings, the green consumer is, if not dead, then at least in a deep sleep." (7)

D004/025/2004

This intention-action gap, manifest in consumers over-reporting their actual green purchasing behaviour, has also been described as "a perceived lack of consumer 'follow through'" on green purchasing claims and intentions (10). The social desirability of being 'seen to be green', and a possible failure to count environmentally-improved 'conventional' products as 'green' products, are suggested as factors here (7). More recent sources detect a continuation of the same scenario into the present (11) (12).

Environmental grounds for purchase are commonly considered to be subordinate to considerations of price and quality. It then follows that environmental factors are generally a source of purchase differentiation only if price and quality criteria are first satisfied (9) (13). On the converse side of the debate, a leading US advocate of green consumption is more confident in contending that the genuinely green consumer is alive and well as a force to be reckoned with in the US marketplace (14). A 1996 study placed the proportion of green consumers in the UK at 10% (6), an estimate typical of its kind. Whatever the precise figure, and whatever the meaningfulness of such estimates, it is safe to assume that the existing green consumer segment can not at present be relied upon to drive more general green consumption in the UK. [1]

3 THE 'RECYCLED CONSUMER'

'Recycled' consumption, i.e., purchase and consumption of recycled products, is one of the most obvious options available to the green consumer.

"Many products contain recycled materials as part of the normal manufacturing process. These include certain types of packaging, such as glass jars, and steel or aluminium cans. There are also products where the buyer can choose a recycled product in preference to one made entirely of virgin material, such as copying paper or kitchen towels. Other products offer an entirely different use for recycled material. Waste plastic is used to make benches, waste tyres to make roofing slates and waste glass to make paving." (16)

Attention has begun to focus on the 'recycled' consumer as a specific class of green consumer. A survey of US manufacturer-vendors of recycled plastic products ranked interest in these products as follows: wholesalers; consumers; retailers (17). While 'consumer' as a class was not defined, it may be assumed that it includes personal and institutional consumers. Personal consumers only are considered in the present paper. Procurement of recycled products by institutional consumers is being encouraged nationally in the UK by the Waste and Resources Action Programme (WRAP), and locally by organisations such as London Remade. This paper seeks to address the more difficult and longer-term challenge of developing consumer markets for recycled products. [2] A selection of empirical studies investigating the personal consumer of recycled products in the UK and beyond are discussed below.

This paper considers explicitly 'recycled' products only, and so excludes products which are recycled but not labelled or acknowledged as such in their consumption. This focus is

---

[1] Typologies of green consumers in the UK have been formulated, e.g, (15).

[2] A paper considering institutional consumers, including business-to-business and procurement purchasing, would be a useful companion to this paper.

intended to directly address the challenge of developing consumer markets for recycled products.

3.1 UK

There is currently a great deal of political emphasis in the UK on recycling collection. Considerable efforts are being made to increase public participation, with a growing body of research into how this may be done. The research into attitudes towards recycling and green consumption identified by the author addresses consumer attitudes towards, and purchase of, recycled products only briefly, however.

A recent overview of existing research into public attitudes towards waste and recycling conducted in the UK, commissioned by the UK Government Cabinet Office [3], states that:

"While the need to recycle ... is well established in the public consciousness, the same is not true of purchasing recycled products. Active purchases of these products remains relatively low ..." (18)

The impressive purview of this report gives weight to its findings. The provenance of the commission [4] also suggests that these findings will inform future waste policy in the UK. Selected findings from some additional and subsequent research not included in MORI's 2002 meta-survey is discussed below. The review of research presented here and throughout this paper is selective rather than comprehensive; its consideration can nevertheless provide the reader with useful and relevant perspective on the current climate for consumption of recycled products in the UK and beyond.

A survey commissioned by Waste Watch investigated attitudes towards waste management and recycling among 1200 members of the UK public, and included participants' reporting of their likely purchasing decisions regarding recycled content products (19). Of the 40% of informants who reported that the fact that a product was made from recycled materials would affect their purchasing decision, an overwhelming majority indicated that this effect would be positive (with certain qualifications, primarily cost and quality), with only 8% reporting a negative effect.

A more recent survey commissioned by the Environment Agency investigated attitudes and behaviour regarding household waste among 2516 consumers in England and Wales (20). A majority (58%) of participants reported finding "no difference" between recycled and non-recycled products on quality. Informants were unclear as to a general price comparison between recycled and non-recycled products, however. A significant proportion (39%) of respondents reported "I am more likely to buy" recycled products than non-recycled.

A report of the findings of a series of 24 focus groups "assessing the public's thoughts on waste as well as their response to the various ways in which it can be promoted" mentions "the purchase of recycled goods" as a component of "responsible shopping" without discussing it in any depth (21).

---

[3] "The review of existing research summarises the findings of more than 20 public opinion surveys undertaken in the past ten years." (18)

[4] "The aim of the survey is to provide the Strategy Unit with an overview of public attitudes towards waste and recycling, to feed into the development of the forthcoming review of the UK Waste Strategy." (18)

Other identified studies of 'green' or 'sustainable' consumption, e.g., (12), do not specifically mention recycled products.

The studies identified above investigate attitudes towards and reported consumption of recycled products collectively, as an abstract product genre. More specific research has focused on household paper products as the most apparent and popular class of recycled products in the UK. A 2002 survey of "attitudes to the environment, and knowledge and behaviour regarding environment issues" was conducted with over 3700 adults in the UK (22). When asked if they had bought recycled toilet rolls or kitchen towels in the last 12 months, 35% of respondents reported that they had done so on a regular basis, 29% once or on a few occasions, and 29% had not done so. An equally recent marketing intelligence report (23) also specifically discusses the status of recycled products in the UK household paper products sector. It concludes that "there remains some reluctance on behalf of the consumer to use recycled products, with many still considering them to be of inferior quality." It is observed that "consumers will only consider buying a product if it is at least as good if not better than the competition and at the same price." [5]

3.2 US
It is generally accepted that recycling policy and practice is more advanced in the US than it is in the UK. There also appears to be more direct research into recycled consumption in the US than in the UK. A number of studies identified by the author which deal exclusively with this topic are discussed below.

In a large-scale poll (24), consumer reaction to recycled products was generally found to be positive: "Almost three out of four (73%) respondents agree that they generally choose products made from recycled materials over products that are not made of recycled materials." In a result which mirrors that for general environmental grounds for purchase (cited above), recycled-content was found to be secondary to considerations of (i) price and (ii) quality, and a source of purchase differentiation only if those two criteria are adequately met. In addition, while 83% of respondents reported that they consider environmental factors when purchasing products, only 15% consider whether the product has been recycled; recycled-content is not widely reported as a green consumption strategy. The same survey also assessed respondent familiarity with the term 'closing the loop'; 'buy / use recycled products' was ranked fourth in the list of meanings attached to the term. It may be possible, however, that respondents were simply unfamiliar with the term itself, rather than the concept behind it.

A survey plotting the attributes of plastic kitchen refuse bags on sale in 35 stores in the US (25) concluded that consumers were unwilling to pay a price premium for the available recycled option. The study suggests this is because the consumer does not directly receive the (environmental) benefit which may result from his/her purchase of the recycled-content product. The incentive to buy recycled is in this sense indirect, and hidden both at the point of purchase and in the subsequent stage of product consumption and use. This contrasts with the direct and personal benefits available to the consumer of organic products, whose motivations have been found to be selfish rather than altruistic (6).

---

[5] This is the only recent UK market research report from this source, found by the author, which specifically discusses recycled products in their own right.

An experimental study (26) investigated consumer preference for identical textile products (sweatshirts and carpet samples) labelled as being made from either virgin or recycled fibres. A minority of the 125 student volunteers (27% (sweatshirts); 19% (carpet samples)) always selected the products labelled as 'recycled', regardless of labelled price. When labelled prices were the same, however, these proportions increased considerably: 66% of respondents selected the 'recycled' sweatshirt and 62% selected the 'recycled' carpet samples. These findings suggest that while a minority will select a recycled product option despite a higher price (given equivalent quality), a majority of consumers will select an explicitly recycled product option if price and quality are the same as a non-recycled equivalent.

A second experimental study (27) investigated the response of 89 student volunteers to two product types (greeting cards, facial tissues), according to two dimensions (recycled / non-recycled, established / non-established brand association). They found that while participants' attitudes towards recycled products were more favourable than those towards non-recycled products, for both product types, this effect was limited to where there was an established brand name association. When the brand name was unknown (in this case fictitious), recycled product status was not found to contribute to brand attitudes. These findings suggest that while 'recycled' products are in general evaluated positively, only those recycled products carrying a known brand name are in a position to exploit this disposition.

3.3 Japan & Germany
A large-scale comparative survey (28) investigated evaluations and consumption of recycled toilet paper by samples of consumers in Japan and Germany.

The two national samples of consumers evaluated a number of suggested ways in which recycling may be improved in their country. Of the six suggested options, increased production of recycled products (54.2% of German sample (ranked 2nd of 6); 35.3% of Japanese sample (6/6)), and increased consumer purchase of recycled products (43.6% of German sample (3/6); 44.8% of Japanese sample (4/6)) were identified by a significant proportion of each sample. These findings suggest a general awareness of the 'closing the loop' concept.

In a blind-test evaluation of virgin and recycled toilet paper products manufactured in the two countries, both sets of participants preferred domestic and virgin products. Within this result, domestic origin was the most important factor for German respondents, while virgin material was the most important factor for Japanese respondents. Respondents were then asked to identify the price condition under which they would use each toilet paper. A clear majority of both sets of respondents reported they would use the recycled papers only if their price was less than that of the paper they normally use.

Self-reporting of toilet paper consumption behaviour showed that "virgin products are consciously used by 35% of Japanese consumers and 41% of German consumers." Conversely, "recycled products are consciously used by 39% of Japanese consumers and 30% of German consumers." There were an array of reported reasons for purchasing virgin and recycled paper; brand was reported as important to purchasers of virgin paper only. Reported consumer perceptions of recycled paper (positive by those who buy it, negative by those who don't) are not explained.

In many studies, consumer evaluation and consumption of recycled products focuses specifically on paper products, and in particular recycled toilet paper; e.g., "[t]oilet paper was adopted as a specific good for study, because consumers have options of buying virgin products and recycled products." Toilet paper appears to be the single-most identifiable, established and visible recycled product, making it an attractive choice for investigation and allowing potential comparison between national markets.

3.4 Summary

The featured empirical studies investigating recycled consumption are summarised in Table 1 at the end of this paper.

4 DISCUSSION

The personal consumer of recycled products has been neglected by research conducted in the UK. The kind of dedicated extensive survey or quasi-experimental studies conducted elsewhere are currently lacking. UK recycling policy and practice has focused on front-end issues relating to collection and reprocessing. The final stage of the 'closing the loop' model, involving subsequent consumption of recycled-content products, has attracted less attention. Markets for recycled products appear to be less developed in the UK than elsewhere. As a result, recycled consumption is currently something of an alien concept to the mainstream UK consumer.

The recent empirical research summarized here nevertheless apparently shows a general consumer preference for recycled products in the UK (as long as they are of equivalent price and quality to non-recycled equivalents), with recycled content reported as an actual incentive to purchase in many cases. This reported preference is not currently manifest in actual product purchase and consumption, however: the 'intention-action' gap. Purchase of recycled consumer products in the UK is at present largely confined to low value items such as refuse bags and paper products. As a result, significant demand for recycled products in the UK must be considered latent at present. We certainly appear to be some way from "the kind of public outcry that prompted legislative mandates for recycled content for a number of plastic products and threatened outright bans of plastic packaging in some parts of the USA in the late 1980s." (29)

The following observation was made in relation to the US context for recycling, but it is perhaps even more applicable here in the UK:

> "[People] appear to love recycling. It seems to meet some deep human need to atone for modern materialism. Unfortunately people do not seem to feel quite the same craving to buy products made of recycled materials." (30)

'Closing the loop' requires that the products made from these recycled materials are then taken up by consumers. For this to happen, a shift is needed in the attitudes of mainstream consumers, for whom recycled products are currently generally seen as an inferior, undesirable option.

Incoming l egislation, p articularly t hat r elating t o the EC Directive on Waste Electrical and Electronic Equipment (WEEE), is likely to promote greater 'closed-loop' recycling in the UK by prompting producers to develop and make available more recycled-content products. Significant claims have been made for the role of legislation in promoting greater recycling in the US:

> "Policy mandates, rather than market forces, have been responsible for the establishment of the plastics recycling industry and have continued to be crucial for its existence ..." (29)

Similar conditions now appear to be emerging in the UK, in the form of demand-side policies focused on expanding markets and increasing demand for recycled materials and the products manufactured with them. [6]

5 DESIGN IMPLICATIONS

Funding and support is increasingly available in the UK, from for example WRAP and London Remade, to designers and producers of recycled products. Within this context, design in i ts v arious f orms c an m ake a number of valuable contributions to closing the intention-action gap in recycled consumption.

5.1 Product design

Good product design is key to developing desirable and competitive recycled products. Recycled products, as a collective class of consumer goods, have a poor image, and are generally considered inferior to what are assumed to be less expensive, better-quality, more stylish and ultimately more desirable non-recycled alternatives. Negative preconceptions around recycled products, and 'green' products in general, will be challenged by the availability of attractive, functional and affordable recycled products. Design has a key role to play here, not just in terms of material substitution (replacing virgin materials with recycled alternatives), but also through more ambitious designing of products specifically with recycled materials. Recycled materials can be superior to non-recycled alternatives; recycled glass, for example, has qualities which make it preferable to virgin glass for certain applications. Recycled materials can also be less expensive than non-recycled alternatives.

There is potential for the exploitation of the aesthetic qualities of products made from waste materials (Figure 2).

---

[6] A recent paper investigates "policy measures that could be introduced to develop markets for such [i.e., environmentally sensitive] products and services." (31)

Figure 2. Re-Form furniture, made from recycled plastic sheeting

"Sustainable consumers need to develop an aesthetic taste for the raw, the partially cooked and the reheated; they need to appreciate the inherent quality in something that once looked like waste." (32)

This issue of a 'recycled aesthetic', and consumer acceptance of products which clearly communicate the non-virgin origin of their materials, relates directly to accounts of the evolution of product design in terms of the functional and aesthetic properties of new materials (33). Rather than seeking to simply pass-off recycled materials in place of virgin alternatives, designers have the more exciting and challenging opportunity to design with, and exploiting the particular qualities of, recycled materials.

5.2 Process design
Innovative manufacturing process design is required to exploit the design and manufacturing potential of any new material. Recycled materials present their own specific challenges in order to be available to designers as a viable option for material selection. The success of the Remarkable recycled pencil, manufactured from one recycled polystyrene vending cup, is built upon the development of a process to convert the recovered source material into a form suitable for (in this case, pencil) manufacture (Figure 3).

Figure 3. Remarkable Pencils Ltd: the Remarkable recycled pencil

Adaptation or 'retro-fitting' of existing production facilities to accommodate recycled materials is likely to be expensive. Nevertheless, as more manufacturing processes are adapted in this way, and as more new production processes are designed to use recyclate rather than virgin feedstock exclusively, so the viability of using recycled materials will increase.

5.3 Communication design

Effective communication design is a key element in the successful marketing and branding of any products, but especially so when those products are 'green' or recycled. Negative preconceptions around recycled products will be challenged if attractive, functional and affordable recycled products are explicitly promoted and marketed as 'recycled'. Opinion currently holds, however, that in order to reach beyond green niches and into the consumer mainstream, green and recycled products should not be promoted exclusively on their environmental credentials.

> "It is vital to stress the direct and tangible benefits provided by greener design, such as energy efficiency or recycled content, rather than stressing the environmental attributes themselves." (13)

Green marketing and branding must be more sophisticated than it has tended to be in the past, and environmentalist messages must be presented to the consumer much more subtly. The survey of research given above concluded that environmental factors are generally a source of purchase differentiation only if price and quality criteria are first satisfied. In that case, appropriate communication design should not then emphasise a product's 'greenness' at the expense of its other more immediately beneficial and desirable attributes (11). Appeals to consumers' personal interest, rather than their sense of altruism, are likely to be more effective in this case.

5.4 Conclusion

In this section, some of the ways in which design in some of its various forms can make a contribution to increasing recycled consumption have been discussed. To be most effective, these would be performed in combination rather than isolation. Highly developed recycled-product branding is likely to be of limited long-term value if those products are of poor or unattractive design or the product manufacturing process is unreliable.

 D004/025/2004

Table 1. Studies of the 'recycled consumer'

	Product type	Method	Participants ▪ number ▪ type ▪ selection
UK			
DEFRA (2002)	Toilet roll & kitchen towel	Survey: personal interviews	▪ 3736 ▪ * ▪ random (stratified)
MORI (2002)	None specified	Literature review: 'meta-survey' of 20 recent surveys	▪ * ▪ * ▪ *
NOP (1999)	None specified	Survey: telephone interviews	▪ 1200 ▪ age 18+ ▪ random (stratified)
Test (2002)	None specified	Survey: telephone interviews	▪ 2516 ▪ heads of household ▪ random (stratified)
US			
Anstine (1997)	Plastic refuse bags	Survey: product data	▪ 35 ▪ retail stores ▪ purposive
King County (1998)	None specified	Survey: personal interviews	▪ 504 ▪ age 18+ / heads of household ▪ *
Mobley et al (1995)	Greeting cards & facial tissues	Quasi-experiment	▪ 89 ▪ students ▪ volunteer
Swinker & Hines (1997)	Sweatshirts & carpets	Quasi-experiment	▪ 125 ▪ students ▪ volunteer
Germany & Japan			
Kishino et al (1999)	Toilet paper	Survey: personal interviews	▪ 998 (G) / 1242 (J) ▪ age 14+ (G) / 18+ (J) ▪ random (G) / random (stratified) (J)

*Note: * = not applicable / unknown*

REFERENCES

1. **DETR** (Department of the Environment, Transport and the Regions) (2000) Waste Strategy 2000 for England and Wales. http://www.defra.gov.uk

2. **DEFRA** (Department for Environment, Food & Rural Affairs) (2003) Municipal waste management statistics 2001/02. (16 September 2003) http://www.defra.gov.uk

3. **CWMre** (Creating Welsh Markets for recyclate) (2003) Recycling loop – overview. http://www.cwmre.org.uk

4. **Harvey, J., Joyce, S.** and **Norman, P.** (2003) Public perceptions of sustainability, energy efficiency and recycling - how these can inform the design process. In: Hon, B. ed. Design and Manufacture for Sustainable Development 2003. London: Professional Engineering Publishing.

5. **Chick, A.** and **Micklethwaite, P.** (2004) 'Closing the loop': lessons from UK recycled product manufacturers. Paper submitted to Design and Manufacture for Sustainable Development Conference, Loughborough, 2004.

6. **McEachern, M. G.** and **McClean, P.** (2002) Organic purchasing motivations and attitudes: are they ethical? International Journal of Consumer Studies, 26, 2, June, pp. 85–92.

7. **Childs, C.** and **Whiting, S.** (1998) Eco-labelling and the green consumer. Bradford: University of Bradford.

8. **Elkington, J.** and **Hailes, J.** (1988) The Green Consumer Guide: From shampoo to champagne, high street shopping for a better environment. London: Gollancz.

9. **Peattie, K.** (1999) Rethinking marketing: shifting to a greener paradigm. In: Charter, M. & Polonsky, M. J. ed. (1999) Greener Marketing: A global perspective on greening marketing practice. Sheffield: Greenleaf.

10. **Peattie, K.** (2002) Marketing: a key gatekeeper for sustainability? In: Charter, M. ed. Managing Sustainable Products: Organisational aspects of product and service development. Towards Sustainable Product Design 7 Conference Proceedings. Farnham: The Centre for Sustainable Design.

11. **Gordon, W.** (2002) Brand Green: Mainstream or forever niche? London: Green Alliance.

12. **Holdsworth, M.** (2003) Green Choice: What Choice? Summary of NCC research into consumer attitudes to sustainable consumption. London: National Consumer Council.

13. **Ottman, J. and Terry, V.** (1998) Strategic marketing of greener products. The Journal of Sustainable Product Design, April, pp. 53-57.

14. **Ottman, J. A.** (2002) The real news about green consuming. The Green Business Letter, May.

15. **NCC** (National Consumer Council) (2002) Green Consuming: Enabling consumers to contribute to environmental improvement. Policy Briefing. London: National Consumer Council.

16. **Waste Watch** (2002) Choosing recycled products. http://www.recycledproducts.org.uk [last updated: 5 December 2002]

17. **Hadjilambrinos, C.** (1999) The USA plastics recycling industry: a survey of manufacturers and vendors of recycled plastic products. Environmental Conservation, 26 (2), pp. 125–135.

18. **MORI** (2002) Public Attitudes Towards Recycling and Waste Management: Quantitative and qualitative review. London: Cabinet Office.

19. **NOP Research Group Ltd** (1999) What People Think About Waste 1999. http://www.wasteonline.org.uk

20. **Test Research** (2002) Household Waste Survey 2002.

21. **Hartley, R.** and **Howes, B.** (2000) Rethinking Rubbish: Towards a new campaign. http://www.nwai.org

22. **DEFRA** (Department for Environment, Food & Rural Affairs) (2002) Survey of Public Attitudes to Quality of Life and to the Environment – 2001. http://www.defra.gov.uk/environment/statistics/pubatt/

23. **MINTEL** (2002) Household Paper Products.

24. **King County Commission for Marketing Recyclable Materials** (1998) Environmental Awareness / Behavior Survey, Summary Report.

25. **Anstine, J.** (1997) Consumers' willingness to pay for recycled content in plastic kitchen garbage bags: a hedonic price approach. Applied Economics Letters, 7, pp 35-39.

26. **Swinker, M. E.** and **Hines, J. D.** (1997) Consumers' selection of textile products made from recycled fibres. Journal of Consumer Studies and Home Economics, 21, pp 307-313.

27. **Mobley, A. S., Painter, T. S., Untch, E. M.** and **Unnava, H. R.** (1995) Consumer evaluation of recycled products. Psychology & Marketing, 12 (3), May, pp. 165-176.

28. **Kishino, H., Hanyu, K., Yamashita, M.** and **Hayashi, C.** (1999) Recycling and consumption in Germany and Japan: a case of toilet paper. Resources, Conservation and Recycling, 26 (3-4), June, pp 189-215.

29. **Hadjilambrinos, C.** (1996) A review of plastics recycling in the USA with policy recommendations. Environmental Conservation, 23 (4), pp. 298–306.

30. **Ackerman, F.** (1997) Why do we recycle?: markets, values, and public policy. New York: Island Press.

31. **Smith, M. T.** (2001) Eco-innovation and market transformation. The Journal of Sustainable Product Design, 1, pp 19-26.

32. **Press, M.** (1996) Crafting a sustainable future from today's waste. co-design, 5/6, pp. 64-69. http://www.co-design.co.uk/co-designindex.htm

33. **Sparke, P.** (1986) An Introduction to Design & Culture in the Twentieth Century. London: Allen & Unwin.

D004/025/2004

Supra-functional factors in sustainable products

D WEIGHTMAN and **D McDONAGH**
School of Art and Design, University of Illinois, Champaign, USA

ABSTRACT

Creating successful sustainable products is as much about making products that fulfil consumer expectation as well as ensuring that they also fit the criteria of sustainability. Outside the hardcore group of eco-warriors, very few consumers will be persuaded to buy products solely because they are sustainable. There is currently no product label equivalent to organic for food products, which might establish this effect. In mainstream industrial/product design it is now recognised that the most successful products are those that combine good functionality with the effective satisfaction of supra-functional needs, including establishing positive emotional relationships between users and the products that they user and buy.

This paper identifies some strategic approaches product design, which can have a significant impact on sustainable product development. These include products which fit a wide variety of user needs more effectively through inclusive design; products designed for repair and refreshment, products customised to fit specific needs; products with long life cycles; retail strategies which maximise the effectiveness of these approaches. Examples will be used to illustrate good practice in these areas with recommendations for design development in the future.

1 WHAT IS SUPRA-FUNCTIONALITY?

Awareness and understanding the way, in which people relate to the products they interact with, is of significant interest to designers. People relate to products in individual and interesting ways. Different people relate to the same product in their own particular way, depending upon its characteristics and their own. Material possessions often serve as symbolic expressions of who we are. The clothes we wear, the household items we buy, and the cars we drive all enable us to express our personality, social standing and wealth (1)(2).

If the user does not perceive a product to be designed for them, barriers may develop. For example, few females would think twice about using a man's razor on themselves. However, few men would feel comfortable using an overtly feminine razor on themselves, though both products perform similar, if not identical tasks.

> ...men and women want to know that the product is "theirs". As evidence, many consumers clearly are psychologically uncomfortable utilizing products and services which do not seem made for them... Therefore, if marketers wish to broaden their

product's appeal across gender lines, they must reposition their product with respect to gender (3).

A study conducted by the authors (4) investigated user attitudes towards electric kettles, televisions and cars. It concentrated on implied product gender and using metaphor to explore peoples' attitudes towards products they used or bought. It became clear that these associations were extremely powerful, especially in the case of cars.

Whilst these aspects clearly apply to mainstream products, do these relationships also apply to sustainable products? And if so, are the relationships different with sustainable, green and environmentally friendly products? Are there particular forms of positive relationship between users and products that can be developed to increase the success of sustainable products?

2 THE LURE OF THE ORGANIC

Inside most supermarkets in the countries of the affluent North, there is an organic section for meat, vegetables and other produce. There will probably be an adjacent section with recycled paper towels, ecological dishwasher tablets, chemical free soaps and various other additive-light and environmentally friendly consumable items. The Body Shop was a pioneer in introducing cosmetic products not tested on animals and produced in an environmentally and socially responsible way.

It cannot be denied that organic products can be commercially viable and are deliberately purchased by some consumers, usually at a price premium. If there was no price premium at all, our suspicion is that the majority of consumers would opt for the organic, ecological or ethically responsible version.

For carnivores, it seems to be reassuring that the chicken you are about to eat had a happy free-ranging life, rather than being a battery slave. Paradoxically the battery chicken may have welcomed the merciful release of its progress towards a McDonald's chicken sandwich, in a way the homespun free-ranger probably would not have. But there is no doubting the smug glow that comes from consumers opting for the celery sticks with real dirt on them, over the standard length ones in the plastic tray.

In other consumer products, there is no equivalent label to 'organic' that resonates in the same way with consumers. In many countries, legislation has made consumers aware of the energy consumption of various products, from refrigerators to cars, but these schemes are usually confined to the publication of information, leaving consumers to make the choice between an economical hatchback and a Chevrolet Suburban Sports Utility Vehicle. Such choices are more complicated by a variety of personal, social, cultural and other economic concerns, many of which will prevail over the concern about energy consumption.

There are significant changes taking place within clothing retail and production. Consumers are being educated about the virtues of organic cotton produced using less harmful pesticides and chemicals, or of clothing fabrics that contain a high percentage of recycled fibres. In keeping with the close affinity of their customers with the environment and concerns about pollution, adventure-clothing companies are often in the vanguard of this approach. In the

USA, Patagonia was an early adopter of polar fleece made from recycled PET soft drinks bottles and the use of organic cotton for recreational clothing.

Cars are, of course more complicated. Arnold Schwarzenegger makes a unique contribution to the future of car design by persuading General Motors to manufacture a civilian version of the United States Military Humvee. At the same time, Tom Cruise is driving around Hollywood in a Toyota Prius, a hybrid petrol-electric car with very low petrol consumption. Apparently he liked it so much he bought two! Was that an ecologically responsible decision or was Tom opting for the nearest thing to "organic" available in the market?

3 WHY DO PEOPLE BUY PRODUCTS?

Before thinking about product development, it is worth giving some thought to why people buy products at all. Individuals purchase and consume products for a variety of reasons. There are many contributing factors that affect purchasing decisions, with a mix of rational and irrationality.

In most cases in Northern countries, these purchases involve people buying replacements for existing products. This may be due to product failure, or the need for uneconomic repairs to existing products. Sometimes it will be to achieve improved functionality, where the existing product has become obsolete or obsolescent, or where the new product performs significantly better. Of course, the reason can also be improved supra-functionality, as products are replaced with ones that look better, have different symbolic associations, are from more highly regarded brands, or are just cooler. As many domestic purchase decisions are not made by individual consumers alone, but as negotiations between partners and family groups, these negotiations can involve combinations of real and spurious functional and supra-functional justifications.

Within the UK, homeownership is relatively high compared to other countries. When individuals rent, they are less likely to invest in consumer products compared to those who own their property (5). Home ownership makes a significant impact on the purchase dynamic. For the new homeowner, identifying, purchasing and installing products within the home environment becomes a time-consuming activity. It is not uncommon for electric kettles to be replaced after two years. As colour schemes change in the kitchen, the kettle may well become obsolete. Colour coordination becomes paramount!

A study found that consumers of small domestic appliances (kettles, toasters and coffee-makers) do not expect such products to last more than two years (6). This is something they are uncomfortable about, but appear accept for the time being.

Some of the authors' eco-warrior friends will advance powerful and convincing arguments that the world already has too many products and too many vested interests working to encourage us to replace them, making them sceptical of the need to replace them at all, even with better products. We tend to take the view that people will replace things anyway, so replacement with products designed and manufactured in a sustainable way is a realistic approach. If these better products do not get replaced with the same frequency as the old ones, then the overall intentions will have been achieved. Not many consumers are the same as a Scottish couple in Parr and Barkers' (7) study of people and their houses, who were bemoaning the fact that their living room carpet (which they hated) was not yet worn out, so it could not yet be replaced, even though it was around 40 years old (refer to figures 1 and 2 with accompanying quotes).

Figure 1: "I remember very clearly when this carpet went down because it came over television that President Kennedy was assassinated." (7)

Figure 2: "But in the 1960s this was really tiptop fashion." (7)

4 SOME STRATEGIES FOR SUSTAINABLE SUPRA-FUNCTIONALITY

All these strategies outlined below are intended to be complimentary to work on product life cycles, energy utilisation, material recycling and all the other approaches which are located in the envelope of sustainable design, with the tacit assumption that all that work is crucial.

4.1 Strategy one: making better products

This approach can be characterised as promoting sustainability by making the product itself better so that user satisfaction is increased, frustration with inadequate or poorly designed products is reduced, so that one of the reasons for product replacements disappear.

"I have the simplest tastes, I am always satisfied with the best." Oscar Wilde

Products that fit their users better will generate positive emotional relationships, will function better and are more likely to last longer in their owners' affections. This is the realm of user-centred and inclusive design where the engagement of users in the design process is vital. This is best achieved by pursuing empathic understanding and participatory design processes.

Inclusive design and user-centred design are two design approaches that place the potential user as the central focus for the product developers. Without fully appreciating the user's experience, aspirations, cognitive and physical abilities, how can a product developer respond in a sensitive and effective way? Such increased sensitivity is going to prove invaluable with the changing demographics and psychographics.

Designers need to go outside their empathic horizon and comfort zone and actively pursue empathic understanding. To expand the designer's understanding and awareness, immersion into the user experience, use environment and wider context helps support evidence based design decision-making. Designers are atypical, not typical users at all. By engaging actual or potential users properly they become a central resource for understanding the context, evaluating concepts and final products.

 D004/010/2004

When we take the time to truly engage people at the level of their personal experiences, such as home, work, play and learning, we are in domains that are truly meaningful to them. We see high levels of creativity in all our participants in such situations (8).

Some designers perceive such involvement as problematic. Users may not be fully aware of their needs and/or able to articulate them adequately, leaving designers frustrated in understanding those needs.

4.2 Customising and evolving products to fit user needs

All users have individual needs, so products should be adaptable to suit those needs. It is essential for the full realisation of the objectives of universal design that product variation satisfies the widest possible user population. This would ensure that the fit between users and products is better for all, rather than providing a good fit for the majority and a continued marginalisation of minority users.

Universal Design is, in fact, about individualisation through diversity, different designs for different users within the same system (9).

Design for adaptability has positive implications for the agenda of sustainable design. As long as the means by which choice is provided is not wasteful of materials or involves excessive design redundancy, the adaptation of products to suit changing user circumstances can be a positive benefit. A user could change product components in response to wear and tear, the availability of components with improved performance or a number of other factors. The use of modular assembly will facilitate this updating enabling people to repair, refresh or even recycle their products, rather than replace them. Fuad-Luke (10) includes encouraging design for modularity as part of his manifesto for *eco-pluralistic design.*

Customisation, whether by designer, manufacturer or users themselves, offers another approach to achieving a better fit between products and user needs of all kinds. Advances in manufacturing techniques have made this truly achievable for mass-market products and so mass customisation has become a reality. This can mean individual specification of options within a mass production system. For example, as when a customer for a new car specifies the engine, trim level, sports package and audio equipments. The modular approach necessary also enables the manufacturer to offer a range of vehicles on the same mechanical platform and using many common sub-assemblies between models.

Of course, wider choice does not in itself guarantee a better fit with user needs. At one stage in its introduction of mass customisation to car manufacture, Nissan had 87 different steering wheels available for its cars. Unfortunately, customers did not favour most of them and were frustrated by the process of finding the one they did want by the proliferation of inappropriate choices. The fact that you can get 250 distinctive covers for your Nokia cell phone does not help people with limited thumb movement or visual impairment to use the tiny buttons or the small screen. Mass customisation can enable a higher level of individual expression, but the design variances must encompass forms and features that are meaningful to the user. Schwartz (11) has found that individuals faced with an excessive number of choices are more likely to decide not to choose at all, rather than agonise endlessly. His study indicates that 6-8 choices appear to be the optimum number of options for most people to cope with.

Customisation and design for variance can also mean design for adaptability, for wear and repair, and for evolution and decay. Adaptability is a means to establish much closer connections between users and products, by enabling them to determine appropriate product characteristics for themselves.

Changing circumstances will affect all product users at different times. Changes might include differences in physical capabilities brought on by illness or aging, increases (or reductions) in available income, changes in living arrangements or living spaces. Products that can be adapted to changing circumstances, to evolve to suit changing user needs, will enable long term positive relationships to be established. One example would be bathrooms designed as wet rooms, so that fixtures can be altered or removed as the users physical capabilities change with age.

4.3 Designing products for longevity

There are few insurmountable problems in designing products for a specific life period, and making them last longer than existing products. A balance needs to be struck between service life, serviceability, and cost of manufacture and purchase price. The situation in which a product is used will determine many aspects of the products life expectancy. If the biggest market for a particular car model is fleet sales, where a vehicle is retained in a company fleet for two years and then re-sold, this will determine servicing strategies, durability requirements, repair and maintenance approaches, and will impact directly on component and sub-assembly design.

Parkard (12) was the first to identify the wasteful nature of planned obsolescence but it is still with us and probably more prevalent than ever. The only consolation is that most modern products work better until they fail.

Some products have useable life spans way in excess of those of their product contemporaries. Porsche sports cars, Ducati and Vincent motorcycles, Airstreams caravans (refer to figure 3), Aga cookers, Dualit toasters, Land Rovers, Barbour jackets – all have durability long beyond the normal life span. Levi 501 jeans used to have the same status, lovingly patched and mended by their owners over a long period of time, but the advent of stonewashing and sophisticated aging processes means that a pre-loved denim state is now available immediately off the shelf. The Airstream caravan was in production in the US from 1936 onwards and approximately 60% of the caravans produced are still on the road (13), cherished by careful owners and sold on via an extremely active second-hand market. That percentage is far above that for even the most collected cars produced in the same period. Such products generate a high degree of brand loyalty and identification, often gaining cult or classic status.

D004/010/2004

Figure 3: Airstream caravan

Cult or classic products are rarely perfect; sometimes their fallibilities are part of their charm of ownership. Often they were expensive as initial purchases, but not always. The combination of functional performance, quirkiness, rarity and brand identification is a powerful mixture, but when achieved, can lead to an extremely strong emotional bond between the owner and the product. Longevity in production also becomes a valuable association.

The diverse mixtures of individuals who make up the Harley Davidson motorcycle owners club in the USA have a shared relationship with a unique and idiosyncratic product range (refer to Figure 4).

Figure 4: Harley Davidson bike

The simultaneous identification with traditional design, American manufacture and the perceived outlaw status of bikers is a potent mixture that people pay large amounts of money to participate in.

> Whisper it please, but the average age of a Harley rider has accelerated from 38 to 46 in the last decade. The best-kept secret of the Harley brand is that its customers are more likely to be accountants and lawyers than unkempt hippies or ferocious Hells Angels (14).

It is still the case that Harley Davidson make more money from selling merchandise, belt buckles, T shirts and other Harleybilia, than from selling motorcycles, but all that will disappear if the bikes do.

Repair, refreshment and recycling can be linked in a coherent strategy for long life products that respond to peoples' needs and facilitate the adaptation and evolution process. The after-sales sector in car customising offers many examples of exploiting modular car manufacturing to enable repair, upgrading and super-tuning. Various engines can be grafted onto the same body platform, sometimes in configurations not envisioned or condoned by the manufacturer, engine components can be changed, and interiors and body panels can be swapped, chopped and channelled. This illustrates how products could be designed for refreshment as well as repair, and this could lengthen the interaction time the owner has with the product.

We are used to refreshment for the biggest product purchase most of us make - property. The costs of moving and relocating mean that this is often the best option for adapting our houses to suit changing needs, both functional and supra-functional. Most houses are not designed to facilitate this process but all houses (and most other buildings) are subject to it. For a full account of the adaptation process, refer to Brand (15).

In most countries of the South, the three R's of repair, refreshment and recycling are linked on a daily basis, as poverty and shortage of materials spawn great inventiveness. One fascinating study by Wolf (16) documented examples in China of people adapting, repairing and resurrecting chairs and other seating devices that we would throw away without any further thought. The following Figure (5) illustrates two fine examples. In case one thinks that this activity is confined to China, the chair below (in Figure 6) was found in a local Korean supermarket in Urbana, Illinois. Note the tennis balls on the end of the legs.

Figure 5: Two examples from the Wolf study (16)   **Figure 6:** Chair discovered in local supermarket

5 Some Case Studies

5.1 Smart Car

The Mercedes Smart car illustrates most of these strategies in action (refer to Figure 7). It is a small two-seater petrol turbo vehicle with a top speed of 80 miles per hour, designed for use primarily, but not exclusively in towns. It has low, but not spectacularly low, fuel consumption. Its length means it can be parked at right angles to the kerb, saving on road parking space.

Figure 7: Smart car

                                                            D004/010/2004

The Smart is available with three engine options, various interior trim options and plastic body panels, which can be relatively easily interchanged or replaced. The manufacturer offers customers an after sales option to change all the body panels if damage or shifting preferences indicate a change. At last enquiry, this was not possible with the interior trim, although it could be and should be. But do not worry, the Internet aftermarket will take care of that. There are a number of German websites for Smart tuning, racing and customising. Up to 125 miles per hour in a straight line, no problem. Passing the elk test, which involves swerving at high speed to avoid an obstacle.

The same structural platform can be altered to produce a soft-top version and has been exploited by the manufacturer to develop a sports car and a four-seater model.

Economical and relatively environmentally friendly in its basic form, capable of adaptation, evolution, repair and refreshment, already a cult object, the Smart car is a good example of a comprehensive approach to creating a sustainable product.

5.2 Naim audio Equipment

Some may be sceptical about citing a hi-fi company as an example of sustainable product design (refer to figure 8). Does this world need high quality audio at all? Naim is a cutting edge audio company that thrives on the particular kind of relationship it establishes with its customers. Even the cheapest gear costs far more than high street hi-fi but its purchase through a small number of specialist dealers brings the purchaser into a close, and supportive relationship with that dealer. Demonstrations precede purchase, part exchange, trade-up, retrofit modifications and a progressive modular approach to system building leads purchasers down a path towards audio nirvana (with its financial consequences). Buyers often stop at points on this path for long periods, not feeling the need to proceed further.

Figure 8: Amplifier by Naim

5.3 Manufactum.de

Manufactum is a German mail order supplier of domestic products including clothes, furniture, tools, household goods, cosmetics, toys and scientific instruments (refer to Figure 9). Long established small companies produce their items all over Europe. Many of the items have been in product for many years, although often forgotten and neglected by mainstream retailers. This is the place to find the badger-hair shaving brush, the wool-felt clogs, the non-drip china teapot and many other fascinating items. These products are not antique or traditional in a nostalgic or sentimental sense, but have been in production for a long time because they work well. Often at a price, but also at the price of not inviting frequent replacement. Those interested in product longevity should study their wares.

MANÚFACTUM.

Figure 9: Images from Manufactum catalogue

6 CONCLUSIONS

In this paper some of the supra-functional factors affecting product design have been discussed. These include connecting people to products, pride of ownership, respect for function, pleasure in performance, emotional bonding with products over a long period of ownership and use. These are elements that accord well with the agenda of sustainable design, but are also elements that relate to the design-process itself.

Better products fit users better and can be designed to evolve or adapt to suit their changing needs. Better sustainable products must do both of those things but also can be repaired, refreshed and recycled. The combination of all five characteristics is the challenge for designers to rise to.

Consumers purchasing fewer products, or keeping their existing products for longer periods will address the agenda of sustainable design. This will not be achieved by legislation but by a number of individual consumer decisions. Those decisions will be based on positive choices of products that are perceived as better. Better for their users, better for the planet.

REFERENCES

1   **Solomon, M. R.** (1983) The role of products as social stimuli: a symbolic interactionism perspective. *Journal of Consumer Research*, 10 (3) pp 319-329.
2   **Dittmar, H.** (1992) *The Social Psychology of Material Possessions: To have is to be.* Hemel Hempstead: Harvester Wheatsheaf.
3   **Milner, L. M.** and **Fodness, D.** (1996) Product gender perceptions: the case of China. *International Marketing Review,* 13(4) pp 40-51.
4   **McDonagh, D.** and **Weightman, D.** (2003) If kettles are from Venus, and televisions are from Mars, where are cars from? In the proceedings of the *5th European Academy of Design conference*, Barcelona, Spain (April), p 151. http://www.ub.edu/5ead/princip5.htm
5   **MINTEL** Marketing Intelligence Report (1996) Small Kitchen Appliances (January) 10.

6   **McDonagh-Philp, D.** and **Denton, H.** (1999) Using focus groups to support the designer in the evaluation of existing products: A case study. *The Design Journal* 2(2) pp 20-31.

7   **Parr, M.** and **Barker, N.** (1988) *Sign of the Times: a portrait of the nation's tastes.* Manchester: Cornerhouse Publications.

8   **Sanders, E., B-N.** and **William, C. T**. (2003) Harnessing Users' Creativity: Ideation and Expression Through Visual Communication. In Langford J and McDonagh D (eds.) (2003) *Focus groups: supporting effective product development.* London: Taylor and Francis.

9   **Steinfeld, E.** and **Mullick, A.** (1997) What it is and isn't. *Innovation: the Quarterly journal of the Industrial Designers Society of America.* (Spring, 1997) pp 14-24.

10 **Fuad-Luke, A.** (2002) *The Eco Design Handbook.* Thames and Hudson: London.

11 **Schwartz, B.** (2004) The Tyranny of Choice. *Scientific American Digital.* April.

12 **Packard, V.** (1964) *The Waste Makers.* Middlesex: Pelican.

13 **Burkhart, B.** and **Hunt, D.** (2000) Airstream: *The History of the Land Yacht.* San Francisco, CA: Chronicle Books.

14 **Harkin, J.** (2003) Brand on the Run. The Independent on Sunday (2 March 2003) pp 28-31.

15 **Brand, S.** (1995)(2nd edition) *How Buildings Learn: What Happens After They're Built. USA: Penguin.*

16 **Wolf, M.** (2002) *Sitting in China.* Göttingen: Steidl.

Product design informed by life-cycle information – initial results from ELIMA

M SIMON and **L ALLMAN**
School of Engineering, Sheffield Hallam University, UK
K YANG
School of Engineering and Technology, De Montfort University, Leicester, UK
T COCK
Merloni Elettrodomestici, Blythe Bridge, UK

ABSTRACT

The EC-funded ELIMA project (2001-5) has developed an information management system which c ollects d ata f rom t wo p rototype p roducts w ith a dditional I T functions. This system records, for example, product operation, patterns of use, temperature and power consumption.

The ELIMA information management system consists of a database with communication facilities, security features to control access by various stakeholders, analysis of raw data with statistical functions, and query and reporting. This system is currently accepting data from batches of two products – a refrigeration appliance and a consumer entertainment product.

In this paper some initial results from the refrigerator field trial are examined. We then ask the question: how can these results be used in design and product development? Examples are given such as the development of test procedures based on accurate knowledge of service conditions.

We present a model of the relationship between the use of life cycle information and the economic and environmental returns, comparing early engineering changes to a just-launched product to improving the performance of the next generation of products. We conclude that the information can be of significant value (exceeding the cost of collection) in reducing future liabilities – e.g. the cost of quality. In addition, the sustainability benefits are potentially valuable – in resource savings, longer life or re-use.

1. INTRODUCTION

This paper describes some of the early results from the field trials of the ELIMA (Environmental life cycle information management and acquisition for consumer p roducts) prototype system for product life cycle data management.

The p roject, w hich i s p art f unded b y t he EC under the FP5 "GROWTH" programme until 2005, aims to develop ways of better managing the life cycles of products by using existing technology to collect operating and usage data such as energy use, time, temperature and shocks throughout a product's lifecycle.

The ELIMA system (Figure 1) provides a database and management software to collect and analyse life cycle information from throughout the life cycle. This information is of great potential value for design, maintenance, logistics, customer service and end-of-life processing (6). Currently, two products containing data acquisition and communication features are being tested in the field trial: a fridge/freezer and a consumer entertainment product. Initial data collected from the field trial and examples of decision making from this data are presented.

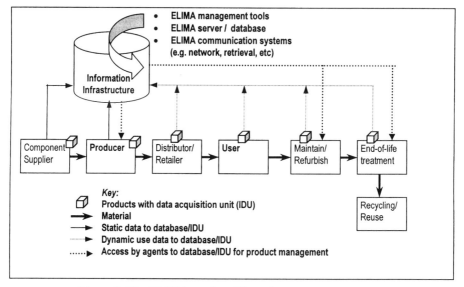

Figure 1. The ELIMA product life cycle management system

2. THE INFORMATION MANAGEMENT SYSTEM

In a typical product life cycle there are requirements for many different stakeholders to access the product life cycle information for decision-making related to the product such as design improvement, associated services or recycling. Such considerations were taken into account during the development of the ELIMA Information Management System (ELIMA-IMS). This is one of the major deliverables within the project, developed to receive and house the product life cycle data acquired by the IDU (Identification Data Unit) (1, 2), and process the data with various statistical tools and data mining techniques to enable and deliver diverse value-added services (e.g., on-line monitoring, fault isolation, etc.) to the stakeholders. In both our prototypes, the IMS is operated by the product manufacturer and in the case of the fridge-freezer, the design and development department "owns" the database and runs queries.

ELIMA-IMS (Figure 2) consists of six modules: the Database Management System (DBMS), the Communication Manager, the Security Manager, the ELIMA Manager, the Information

Engine and a web-browser based interface. This modular design ensures the reusability, maintainability and scalability of the system. In order to aid the building of the system, a combination of development tools and a software framework consisting of a set of components, class libraries and a life cycle database schema have been developed (3). This will help the developer to rapidly build a product-specific ELIMA-IMS. Two prototype systems have been developed using supporting tools and the ELIMA-IMS framework to support the field trials and are currently up and running at the Intelligent Machines and Automation Systems (IMAS) laboratory in De Montfort University. The main features of the prototypes are as follows:

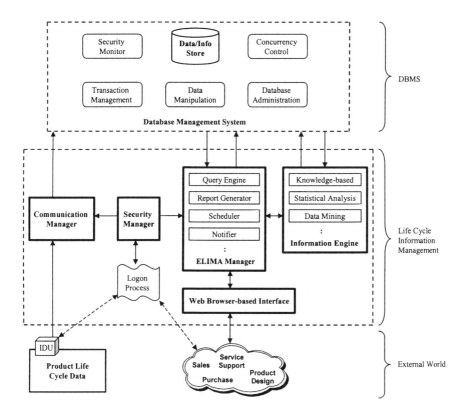

Figure 2. The ELIMA Information Management System

• the run-time Infrastructure supports multiple databases (e.g. for fridge/freezer, for games console, etc) running at one single server without interrupting each other.
• the Communication Manager supports various product-specific life cycle data transmissions, e.g. via internet file transfer.
• the DBMS supports various database management systems (e.g., MySQL, Oracle, etc).
• the Security Manager supports password access and customisable views and facilities for different stakeholders.

- the ELIMA Manager at the core of the system includes the functions of making queries, generating reports, notifying events (such as data arrival) and scheduling regular actions.
- XML is used as a neutral language for receipt of product data "packets".
- there is a facility to allow an administrator to manage the database.

In practice, users make basic queries using the browser interface and then download selected data files for off-line analysis.

3 EXAMPLES OF LIFE CYCLE INFORMATION

The system processes raw data from the products into usable information which can support design and development, as well as other management activities such as logistics, maintenance planning and end-of-life processing. In the current project, refrigeration products from the initial production runs under long-term test are providing valuable information in three areas:

- Design – information on product performance in users' homes, such as energy consumption, can validate designs and inform future design; for example, automatic defrost frequency varies significantly between users;
- Marketing & user behaviour: deeper understanding on customers' habits can be gained, for example the variation in fridge door opening times (see Figure 3 – where surprisingly high variation in user behaviour is evident that can be correlated with energy use);
- Policy and standards: user test information can confirm the suitability of standard tests such as the energy label, and inform company approaches to standardisation discussions, as well as UK or EC policy on domestic energy consumption.

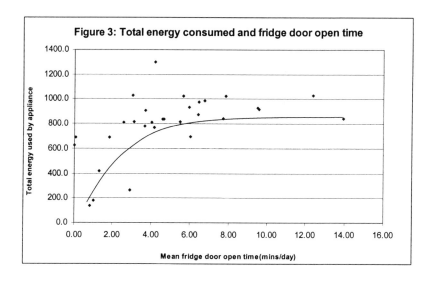

Figure 3. Example of processed data

D004/008/2004

However, the data processing effort required to extract usable information should not be underestimated. The ELIMA project is evaluating the typical costs of this activity.

The other ELIMA field trial is giving data relevant to repair and refurbishment of failed products. When the consumer entertainment products are returned by the user for repair, the IDU data is extracted by an RFID interface even if the product is powered off. Rapid decision-making at this stage cuts the cost of repair - products that will not be economic to repair can be recycled. The remaining life of some parts may be predicted more accurately – this is the subject of ongoing research.

4. USING LIFE CYCLE INFORMATION IN DESIGN AND DEVELOPMENT

There is no doubt that the quality and level of detail in the information collected so far vastly exceeds that which was available from previous methods such as lab tests and user questionnaires. It is exciting for a designer to be able to see how real people actually use their products, often within hours of the events recorded. The information can be considered as of two types: *operation*, which relates to whether the product functions as designed under known conditions, and *usage*, which is data on the actual environmental conditions (e.g. ambient temperature, shock) and user behaviour.

So how will the use of life cycle information affect product design? How will future products differ? We can currently see at least six design-related applications:

1. Performance, cost and durability – the ELIMA system gives what is effectively an extended, or even open-ended, field trial of a product design. The data on whether the product is operating as the designers expected allows better evaluation of component performance and life in real conditions, leading to better and more durable future designs. (*operation and usage*)
2. Interaction and control – the data on user behaviour as they interact with the product gives better understanding of users than limited tests and trials. For example, the fridge/freezer under test has a "holiday" setting to reduce energy consumption when the user is away. How often is this function used? (*usage*)
3. Test procedure design – accurate knowledge of user behaviour allows test specifications to be refined. For example, our door opening frequency data was used to inform a test of freezer drawer durability. (*usage*)
4. Energy labelling – results on product performance show whether the "standard" tests for energy labels are realistic. Are users getting the promised savings from their "A" rated product? Ideally, the product design should be tuned to give a good performance in the standard test and similar low consumption under a wide spectrum of actual use conditions. (*operation*)
5. Maintenance a nd r eliability – i n future products, features can be planned which are cost-effective ways of aiding fault diagnosis and repair. Understanding the interaction between standard components in operation, such as the effect of temperature and vibration on circuit boards, helps design for maintenance; the long-term performance of seals and insulation materials, which slowly degrade over time, is also hard to test otherwise – but affects the energy consumption of the appliance. (*operation*)

6. End-of-life – the simplest piece of data, the product age, is usually unknown by organisations refurbishing household appliances. Access to information about disposed products will promote their reuse; parts can also be refurbished or reused if data is good enough to estimate remaining life, a subject of ongoing research in the ELIMA project. The complementary design activity involves anticipating multiple lives for products or components and designing in appropriate features, such as separate IDU devices in motors (4). (*usage*)

Our initial assessment of the information is that its value depends on its accuracy and timeliness. In our prototype systems, considerable effort has been required to validate the data transmitted from some of the products, especially when large quantities of data are being obtained. (Some of the fridge/freezers under test send 128 kilobytes of data every day by the mobile phone – internet link.) Validation is important before the data is processed (e.g. by averaging) to avoid false readings distorting the conclusions.

Simulation of the product life cycle is also useful, since reliability is a critical factor in design. Quality depends on achieving a level of reliability acceptable to the user and this is expensive – but the cost of faults, both in repairs under guarantee and in loss of reputation, may be even more costly. We have developed a complex model of the entire product life cycle, using Arena software, that allows the effects of reliability to be simulated. Adding ELIMA functions to the product can then be evaluated in terms of costs and benefits.

The timeliness of the information is relevant because of the logic of the product design cycle. Figure 4 illustrates the timescale of information feedback into the design cycle. From a field trial of prototypes or pre-production models, designers get feedback early enough to make engineering changes on the same product or range. The longer-term feedback of information from the mainstream product sales is only useful for the design of the next product or range. Finally, and not shown in the diagram, collecting data at end-of-life is probably too late for design but may assist the recovery of value from an individual product.

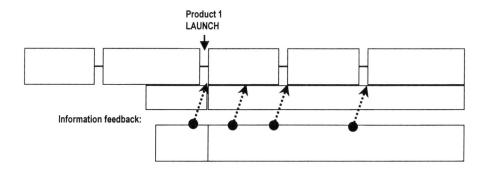

Figure 4. Timing of information feedback to design

For mass-produced consumer electrical and electronic goods, the design cycle can be as short as six months or less for entertainment products or mobile phones. However, it may be three to five years for white goods, with major retooling on an even longer cycle (5). Once a product has been launched, life cycle information coming back is only good for design of the next model – and may in fact be relevant to the next generation, several product launches in the future. The only exception is that use of life cycle information from "beta-tests" or pre-launch field trials can help with engineering c hanges i n t he i mmediate p ost-launch p eriod. These changes, which are usually invisible to the purchaser, would fall into the first two categories of design application above. Engineering changes may involve software modifications to refine performance, components from a different supplier or of a different cost, or manufacturing procedures.

In the two cases detailed above where feedback is useful for design, it is not essential to have data on every product – a sample will do. This can reduce costs: either a special add-on IDU is inserted into a sample of early products, or all products have the feature but data is only collected from a selection.

5. CONCLUSIONS

For effective product life cycle management, an information system is likely to be operated by the producer and needs to possess a sophisticated database with access control and security features. Data collection and processing is a significant cost overhead, since validation is important and data volumes can be high.

However, if a system is operated where data can be collected early enough to feed back into design, the cost of quality can be reduced – in other words, the net benefit is a reduction in producers costs and an improvement in product quality. Future products will be a closer fit to users' needs and will have better environmental performance as well as increased life. This is likely to lead to more sustainable products, even without any changes to end-of-life practice; the additional benefit of the life cycle data for end-of-life decision-making is not directly relevant to design.

REFERENCES

1. **Moore, P. R., Pu, J., Xie, C., Simon, M., Bee, G.** (2000) *Life Cycle Data Acquisition Methods and Devices*, Proceedings of the 7th Mechatronics Forum International Conference, Mechatronics 2000, 6th – 8th September 2000, Atlanta, Georgia, USA.

2. **Simon, M., Moore, P.R., Pu, J.** (2001) *Modelling of the life cycle of products with data acquisition features, Computers in Industry* **45**, 111-122.

3. **Moore, P. R., Pu, J., Wong, C. B., Chong, S. K., Yang, X.** (2003), *A Component-based Development Environment for Environmental Life-cycle Information Management Systems for Consumer Products*, International Conference on Computer, Communication and Control Technologies, CCCT 2003, July 31, August 1-2, 2003, Orlando, Florida, USA

4. http://www.dyson.co.uk/news/article.asp?mode=Back&id=130, last visited 11/5/04

5. **Open University**, (1997). *Green product development: the Hoover New Wave.* Video and coursebook, Design Environment and Strategy T302(VC2).

6. **Simon, M & D ixon, A.** (2003) *Product Life Cycle Management for Sustainability through Information Technology*, in Design and Manufacture for Sustainable Development 2003, Cambridge, September, pub. PEP, London, 159-168.

D004/008/2004

Strategies Towards
Factor 10

EcoDesign and future environmental impacts

J JESWIET
Visiting Fellow, Australian National University, Canberra and Queen's University, Kingston, Ontario, Canada
M HAUSCHILD
Danish Technical University, Lyngby, Denmark

ABSTRACT

This paper describes the relation between EcoDesign and Life Cycle Engineering; both include Product Engineering as a focal point. Product Engineering includes both Product Design and Manufacture, two fields which are changing quickly. In addition, this paper shows where future changes can be expected in Life Cycle Engineering and EcoDesign.

1 INTRODUCTION

EcoDesign is a new concept with Product Design at its core. It is one of the terms being used to describe the approach used by Product Designers in the multi-faceted field of Product Design and Manufacture. The life of a product starts with the initial design concept and getting the design right, at the beginning of a product's life, is most important to the eventual cost of the product, as shown by Boothroyd and Dewhurst (1). They found that seventy percent of the final cost of a product is determined at the design phase. This can be extended to functional requirements and environmental impacts. The impact of a product upon the environment is determined at the design phase, hence the importance of EcoDesign and associated concepts such as Life Cycle Engineering, LCE.

Predicting where future technological changes will occur is difficult, if not impossible at best. Predicting where potential, future environmental impacts will be is even more difficult, however with the advent of some of new technologies it is possible to make some reasonable predictions. This paper looks at the relation between EcoDesign and LCE and tries to predict where future impact areas will be.

2 LCE and ECODESIGN

In the work by Lagerstedt (2), on "Functional and environmental factors in early phases of product development Eco Functional Matrix", it is stated that "Design for the Environment is also known by numerous other names such as: Green Design, Eco-Design, Sustainable Design, Environmental Conscious Design, Life Cycle Design, Life Cycle Engineering and even Clean Design". Lagerstedt goes on to say that although the wording may have different meanings, the terms generally have the same goal. This is true, however there are some differences, although minor ones in some cases.

Life Cycle Engineering is the aegis under which many current approaches to environmental work can be placed. The question, what is Life Cycle Engineering? was posed to colleagues involved in LCE and the result (3) gave similar yet diverse definitions. The following gives a general definition, with keywords.

In trying to define LCE, reference was first made to the Handbook of Life Cycle Engineering, but to no avail. Surprisingly there was not a definition in a book dedicated to LCE (4). After subsequent, global consultation with colleagues, it was found that the diversity of Engineering Environmental work falls under the aegis of Life Cycle Engineering, or LCE. LCE is a conglomeration, which includes many subjects, it is Engineering work, which includes Product Design and the Environment. A definition (3) of LCE is:

Engineering activities which include: the application of technological and scientific principles to the design and manufacture of products, with the goal of protecting the environment and conserving resources, while encouraging economic progress, keeping in mind the need for sustainability, and at the same time optimizing the product life cycle and minimizing pollution and waste.

The keywords of LCE include those shown in figure 1. LCE includes the many aspects of EcoDesign and in many cases the two terms can be used interchangeably. Differences in EcoDesign and LCE are that LCE includes in the service industry, Life Cycle Assessment, LCA, and Product Manufacturing.

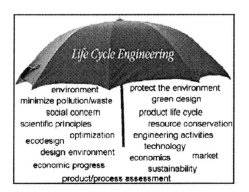

Figure 1. Keywords of Life Cycle Engineering (3).

3 LCA and PRODUCT DESIGN

Life Cycle Assessment, LCA, a subset of LCE, is generally accepted as a method of conducting an assessment of environmental impacts. LCA is used for evaluating environmental impacts associated with either a product or a service. An LCA includes the entire life cycle of a product, from cradle to grave: extraction of raw materials, processing, manufacturing, transportation, distribution, use, maintenance, reuse, recycling, disposal. SETAC (5) gives a definition of LCA as:

'An objective process to evaluate the environmental burdens associated with a product, process, or activity by identifying and quantifying energy and material usage and environmental releases, to assess the impact of those energy and material uses and releases on the environment, and to evaluate and implement opportunities to effect environmental improvements.'

The Design Phase of a product is the most important phase in the life of a product, especially from the point of view of functionality, cost and the environment. Decisions made at the design stage remain with a product until it's End-of-Life. The decisions made before manufacture will have large environmental impacts and must be supported by an LCA as early as possible, if the product is to have minimal impact on the environment.

An important aspect of LCA is the product End-of-Life, or EOL. The product EOL must be dealt with according to jurisdictional requirements, and this is becoming increasing important, as c an b e s een w ith d irectives s uch as the EU directive for automobiles (6). Feldmann (7) discusses EOL in terms of waves of product return ratio, and show the need for strategies that must be applied to Product Design in order to deal with product EOL; See figure 2.

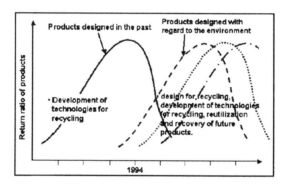

Figure 2. Return ratio of products in terms of waves as predicted by Feldmann (7).

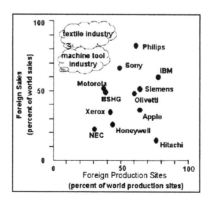

Figure 3. The trend toward Globalization positioning of major industries (7).

Feldmann (7) goes on to show that in many situations there is a global relocation of assembly industries, specifically to take advantage of lower wages in less developed countries; see figure 3. However, this is not the case for all industries, such as the machine tool industry. The machine tool industry is a special case in which production lot sizes and product complexity requires a close cooperation between design, assembly and startup as well as highly skilled workers, especially at the assembly stage. However, the globalization of markets, as well as technical innovations, increasingly cause a redesign of the arrangement of the value adding chain in the global manufacturing network. Assembly either follows the markets and/or is shifted into regions of lower costs.

4 CHANGING TECHNOLOGY and ECODESIGN

History is about change, and observing those changes can help in predicting where we are headed. Predicting where future technological changes will occur is difficult, if not impossible at best. Predicting where potential, future environmental impacts will be is even more difficult, however with the advent of some of new technologies it is possible to make some reasonable predictions.

The changes in technology that have occurred since the industrial revolution, and the evolution of technology, can be visualized as waves progressing over time. Sheng (8) proposed a model, which shows how the period for technological waves decreases over time. Part of this model is shown in figure 4. This model can be expanded to include the next wave, as shown, and it implies that we are in the fifth wave and that we are at the peak of that wave. The implication is the next wave will include Engineered Earth Systems and Biotechnology, and these likely will start around 2020.

The wave model implies the period of each wave has continually decreased and that, with some analysis, it can be shown the rate at which change can be expected to occur will reach an asymptote in the next decades.

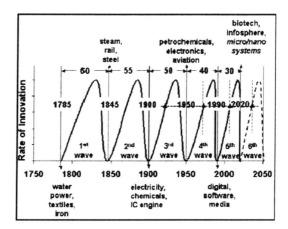

Figure 4. A model of the evolution of technology can be shown in terms of waves.

                                                                      D004/015/2004

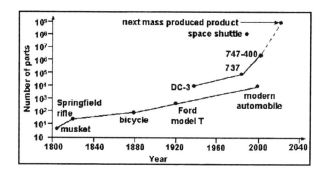

Figure 5. The increase of the number of parts used in a product, as manufacturing methods have increased in sophistication.

It may also be observed that industrial products have become more complex over time (8). Figure 5 shows how the number of parts included in a product has increased exponentially, since the start of the industrial revolution with the introduction of newer manufacturing techniques, especially automation of manufacturing processes. It can be seen that product complexity has increased and the period for each successive wave has decreased.

With shortening lead times, and more complex products, which may have different functions, designers must be faster in identifying potential environmental problems and there is need 'to get the focus right' at the intersection of the design process and the impacts upon the environment.

The foregoing also implies the rate of increase in innovation has increased. Also, there is a ramp-up time, for instance, biotech products have been in the market place for awhile, whereas the wave model shows they are just starting. This means there is a space in which there is time to react to potential, future environmental impacts of a product.

Micro/Nano Technology have been included in figures 4 and 5 because there is intense activity in this area at present, for example the research activities outlined in a recent CIRP seminar (9). This is the 'new sixth wave', which is not included in previous model by Sheng (8). However, it will have a dramatic effect upon future designs, and by extension, product function and the environment. Using the foregoing model for innovation, it may be seen that we are in the ramp-up phase, and in addition, that miniaturization will have an effect, possibly cause a refocusing of product function with a subsequent effect upon Design for Environment, and by extension EcoDesign.

5 THE IMPACT OF AUTOMATION

As micro-manufacturing applications increase, there will be an increase in the number of parts per p roduct. It c an b e predicted that automation will be involved, because micro and nano manufacturing cannot be accomplished without techniques used in automation, including new techniques being developed to for manufacturing purposes. Hence there will be increasing productivity with a concomitant rise in hard automation and eventually in the number of robots used.

It can be predicted, using the information provided by Feldmann (7), that sectors parts of the machine tool industry will be involved, but complex, bulky sectors will likely not move to a new location. However, a robot is, in effect a machine tool, and when micro-manufacturing becomes increasingly prevalent there should be a rise in the use of robotic devices in manufacturing. Since robots are portable, their use is not physically rooted. Figure 6 shows how robot usage now stands. It can be seen that new installations have not increased recently, but this can be expected to change in the future, especially with increased activity in micro/nano manufacturing. An increase in robot use should be a future indicator of increased micro/nano manufacturing.

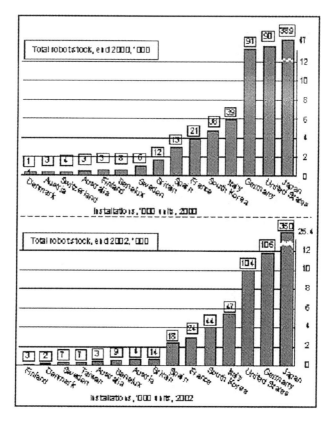

Figure 6. Robot usage for 2000 and 2002 (10).

Any final product is intended for the marketplace, which inevitably means human beings. The size of human beings is constant, hence the products they use will match their size. Although there is a need to produce smaller products, such as smaller, more effective hearing aids and other Biomedical devices, not all products will be for the Biotechnology market. Many products will be on a 'human scale'. With the drive toward miniaturization of products, it can be seen that both Micro and Nano Technology will likely contribute to an increase in the number of parts included in a product, and that they will also be part of the 'sixth wave', driving the development of Biotechnology and Engineered Earth systems.

Products have also increased in complexity, as was shown in figure 2, and with the increase in miniaturization there will be increased variability in a product, making products even more complex and, inevitably, increasing the number of functions possible, or even changing the function of a product. Therefore miniaturization may not decrease environmental impact, but may act in reverse increasing environmental impacts as product function changes emphasis, becomes more complex, and requires more innovative methods of production. The methods of production developed will likely have environmental impact of which the EcoDesigner must take into account. There will also be disassembly issues, End-of-Life and take-back issues of which the EcoDesigner must be cognizant.

Environmental Impacts can be reduced if an LCA is done early, at the design stage, which is in effect, "getting the focus right" in product Life Cycle Assessment (12). Getting the overall focus right in Product Design with an eye to Disassembly, Recycling and (re)Manufacturing will be even more critical as products become more complex. Products must not only be micro-manufactured but it must also be possible to do micro-disassembly and to take EOL into account. Hence, Micro manufactured products must addressed from a product sustainability point of view. This calls for overall life cycle optimisation of industrial activities, which include (12):

- Products, which are at the core of environmental impacts,

- People, profit and planet which all impact the environment,

- Product Design to reduce overall environmental impacts.

This means addressing the whole product life cycle, hence product development must rely on LCA as analytical tool, and these must be applied to new processes and products, which are increasingly more complex.

6 ENVIRONMENTALLY BENIGN MANUFACTURING

The EcoDesigner must also be cognizant of changes throughout the manufacturing sector, especially with respect to the environment, and at the international level. This includes activities under the heading of Environmentally Benign Manufacturing, EBM.

In 2000, the National Science Foundation conducted a study of manufacturing and it's effect upon the environment (9). The primary purpose of the study was to survey environmental work in the manufacturing sector in other countries, and to define objectives for similar, future work in the United States.

The focus areas were:
- Metal Processing
- Polymer Processing
  - thermoplastics
  - thermosets,
  - composites
- With applications to
  - automobiles
  - electronics

The subsequent report (9) addressed:
- development and implementation of benign materials processing to meet the challenges of sustainable materials flow in a use and reuse environment, and
- the systems consideration of re-manufacturing, reuse, and recycling in total waste-stream management.

In this study, the group preferred to use the term Environmentally Benign Manufacturing, EBM. The rationale for this appeared to be that Sustainability was overused and too fashionable. They defined EBM as:

- enabling economic progress while minimizing pollution and waste, and conserving resources.
- protecting the environment for the next generation; this is a tenet of sustainability.

In addition, the EBM study listed the following future needs (9):

- products should be designed for re-use,
- better reprocessing technologies must be developed,
- EBM is "lean" rather than new,
- integration of financial and environmental systems,
- re-use / life prediction modelling,
- accounting system for the "value" of EBM in processing/design selection.

It may be observed that EBM places the emphasis not only upon Manufacturing, but it also recognizes that Design is extremely important and that doing a proper EcoDesign will decrease the Environmental impact of a product, before it even gets to market. In other words, a proper EcoDesign will take into account the use phase, and End-of-Life requirements.

One difference that can be observed in EBM is in the association of the word progress with economics, and the future need for the "integration of financial and environmental systems ". This infers a direct link between environmental progress and financial progress. With respect to this, other information can be taken into account. For instance, a recent interview on the Australian Broadcasting Corporation (12) indicated that Banks were finding that companies, that were managed in a more environmentally responsible way, were more likely to be better managed financially, with the result that the same companies had far better financial positions. One company, formerly a large producer of Greenhouse gases, was shown to have gone ahead and dealt with the Kyoto accord requirements for reduction of Greenhouse gases, in-house.

The result was their balance sheets showed a net gain of 650 million dollars, after Greenhouse emissions had been reduced throughout the company. Based upon the foregoing, the prediction is being made (12), that in future, Banks will make higher interest loans to companies, which have a poor environmental, in-house, ethos, because of potentially higher financial risks; in other words a poor environmental ethos is a strong indicator of a poor financial management with concomitant higher financial risks.

7 INDICATORS IN THE CORPORATE SECTOR

The example of positive action in the foregoing section, by a company deciding to meet the requirements of the Kyoto agreement on greenhouse gasses, is not the only example. Another example of how companies have become concerned about their environmental image and how this will have an impact upon sales of their brand is given by the report by Macalister (13) that an oil company has been holding secret meetings with environmental groups worldwide in an effort to change its hard-nosed public image. Critics claim the company has played a major role in the fight against the Kyoto treaty on climate change. According to the article competitors have a softer public relations image.

It can be predicted, that companies not only need to change their image in many cases, but there is a need for tools which can be used to assess both their old and new product lines, and whether a new area of endeavour can be both profitable and environmentally responsible. Part of that assessment will require an Environmental Risk Analysis, which will probably be included in the tools used by the EcoDesigner.

8 CONCLUSION

Changes can be expected due to micro/nano technology with concomitant environmental impacts. Changes can also be expected with the increasing complexity of products and in the machine tool and automation tool sector with the development of new techniques. The increased rates of production needed and, possibly, increased environmental impacts.

In addition, jurisdictional rules will be applied even more in order to control environmental impacts. These factors should be taken into account by the EcoDesigner.

New EcoDesign tools will be needed at an early stage to identify functional, economic or environmental problems and any associated risks. Risk analysis will become more prevalent. In addition, conducting an LCA, as early as possible, will also become increasingly important and may even become a product requirement in some jurisdictions.

REFERENCES

1. Product Design for Manufacture and Assembly. Boothroyd, Dewhurst and Knight. © 1994 Marcel Dekker, NY.

2. Lagerstedt, J. KTH, January 24, 2003 "Functional and environmental factors in early phases of product development -Eco Functional Matrix", PhD thesis.

3. Jeswiet, J. "A Definition for Life Cycle Engineering". 36th International seminar on Manufacturing Systems. June 3, 2003. Saarbrucken Germany. Plenary Speech. See proceedings pp 17 - 20.

4. Life Cycle Engineering Handbook, © 1998 Elsevier.

5. SETAC, Society of Environmental Toxicology and Chemistry; Pensacola, Florida, USA.; http://www.setac.org/.

6. EU automobile take back directive; Directive 2000/53/EC, 18 September 2000, on the EOL of vehicles.

7. K. Feldmann, H. Rottbauer, N. Roth, 'Relevance of Assembly in Global Manufacturing', Keynote paper, 1996 Annals of CIRP vol 45/2/1996; pp 545 - 552.

8. Graedel, T.E. and Allenby, B.R. Industrial Ecology. © 2003 by AT&T, published by Prentice Hall, New Jersey.

9. Workshop on Environmentally Benign Manufacturing. National Science Foundation the Department of Energy. July 13, 2000.

10. Economist. October 18th 2003, page 98.

11. Hauschld, M. Jeswiet, J. Alting, L. 'Design for the Environment - Do We get the Focus Right?' Accepted for publication in the Annals of CIRP, vol 53/1/2004.

12. Australian Broadcasting Corporation radio interview, 0815, June 1, 2004.

13. "Exxon seeks to clean up is image as global villain". T. Macalister. Guardian Weekly, October 16 - 22, 2003, page 7.

D004/015/2004

Understanding the potential opportunities provided by service-orientated concepts to improve resource productivity

M COOK
Institute of Water and Environment, Cranfield University, Bedford, UK

ABSTRACT

Driven by the need to secure additional economic value, manufacturing firms have integrated a variety of service orientated concepts into their product portfolios. In some instances these are additional to their existing traditional material based products whilst in others they supplant these altogether. A number of commentators contend that this emerging technological trajectory may provide a promising avenue toward ecological modernisation and opportunities to yield significant improvements in resource productivity.

Using the results of a stream of research conducted by Cranfield University to assess the opportunities that this trajectory may provide, this paper identifies and challenges many of the key assumptions which underpin this assertion. More specifically, it uses the results of case study research to provide in-depth insights into instances where Eco-efficient service orientated concepts have been transferred to and applied within traditional manufacturing companies, in order to identify opportunities to improve resource productivity and importantly, how these might be realised.

1 INTRODUCTION

Service now accounts for approximately two thirds of western economies and instead of purchasing material products, consumers and businesses are increasingly buying a range of services: cleaning services instead of washing machines, document services instead of photocopiers, voicemail instead of answer phones. In this new service or functional mode of consumption, the focus of activities is not goods in and of themselves but rather the service or utility which these deliver. Thus whilst material products were formerly the focus of attention in a material based economy, in a service economy these are merely artefacts used to support service offerings.

Several policy discourses suggest that this trend in western economies offers great potential to improve the environmental performance of production and consumption systems. These are concerned with industrial ecology, ecological modernisation and eco-design, and respectively emphasise the potential of services in the attainment of dematerialisation, development of improved eco-efficiencies and greater integration of eco-design principles and practices.

Research undertaken at Cranfield University has sought to investigate these discourses and their claims within the UK context and particularly the UK Government's intention to create sustainable patterns of production and consumption. Whilst it is not possible to report the entirety of this work, emphasis has been given to investigation of services within the context of ecological modernisation and the need to improve eco-efficiencies and associated technical research undertaken by those concerned with Eco-design.

The paper is necessarily divided into three sections. The first provides an overview of work to date; the second reports research undertaken by Cranfield University in the UK to explore the potential of services to improve environmental performance in the UK; and, the last draws some conclusions from the findings of this work.

2 OVERVIEW OF WORK TO DATE

Not all services are eco-efficient or indeed, in the widest sense, sustainable. Thus the intention of research has not been to facilitate the unfettered development of services but rather to assess the opportunities that this shift in consumption practices provides to increase resource efficiencies and to provide new product concepts and support tools to assist with this matter. Synthesis of work completed to date shows that research has focused on three main areas:

- New Product Service Systems
- Development of New Product Service Systems
- Marketing and Consumer Acceptance of New Product Service Systems

2.1 New Product Service Systems
Whilst the rapid expansion of service sectors has been partly fuelled by an increase in traditional services (e.g. banking, insurance, law) a range of new service orientated products have emerged in sectors where demand was traditionally satisfied through the provision of material products. A number of these are detailed below in table 1.

Table 1 New Service Orientated Products

Service	Description
Chemical Management Services (Castrol Industrial North America)	Manages chemical procurement, delivery, inspection, inventory, storage, labelling and disposal for industrial customers. Seeks process efficiency improvements. Compensation can be based on cost savings delivered, not volume sold.
Document Services (Xerox)	Integrates document storage and reproduction technology – Xerox's traditional manufacturing strength – with customer's business systems to produce automated, just in time, customised document production.
Mobility Services (Call a car Netherlands)	On demand car rental. A fleet of cars is owned by a membership organisation; subscribers pay fixed costs and per kilometre/per hour fees. Cars are reserved "on demand" via a central reservation point.
Furnishing Services (Interface; DuPont Flooring Systems)	Interface experimented with an "Evergreen Lease" program. Customers leased installed modular carpet, which interface undertook to maintain to a given appearance standard with selective rotation or replacement (with recycling) of worn tiles. DuPont, in addition to leasing carpets, also provides a series of carpet related services throughout the carpets lifecycle.

D004/013/2004

New service offerings have been found in both business to business and business to consumer markets. These range from air conditioning companies providing 'Coolth' instead of air conditioning units to the provision of extended warrantees which effectively insure the performance of 'white goods' throughout their design life. This qualitative shift in patterns of production and consumption has been conceptualised by many as the 'shift from products to services (Cooper and Evans, 2000; Mont, 1999). Whilst there are some issues concerning nomenclature (services are products and services often need material products to support delivery), this has been the locus of product based research.

The results serve to illustrate that the service type of transaction is finding increasing popularity in sectors of industrial economies where the material product based transaction formally prevailed. Research has focused on analysis of product service systems, which is predicated on the notion that a product service system consists of tangible products and intangible services designed and combined so that they are jointly capable of fulfilling specific customers needs (Tischner et al, 2001). It also shows that whilst artefacts are a key component of new product service systems, they are often subordinate to the service element. Thus it is understood that manufacturing firms are becoming increasingly focused on providing results not 'products', whilst customers are becoming increasingly amenable to consuming results. From this research, a spectrum of new product service systems have been found and include:

- Product Orientated PSS

Product transfers into the ownership of the consumer but additional services are provided, e.g. maintenance contract.

- Use Orientated PSS

Product is owned by the service provider, who sells the functions instead of products, by means of modified distribution and payment systems, e.g. sharing pooling and leasing systems

- Result Orientated PSS

Products are substituted by new services, often driven by new technologies, e.g. voicemail replacing answer phones.

In response, a number of authors argue that the emergence of these new service orientated products will provide opportunities to significantly improve the eco-efficiency (by factor 4 and above) of production and consumption systems (Brezet et al., 2000; Giarini, 1987; Manzini,1996; Meijkamp, 2000; Ottman, 1998; White et al., 1999). These claims are based on the notion that by focusing on a product's functions, new sometimes radical ways of transforming what might be described as the 'product service mix' can be found that satisfy demands whilst simultaneously improving environmental performance. And importantly, that these can be realised through the application of certain product concepts. These include, *inter alia*, the notion of Sustainable Product Service Systems (PSS) and Eco-Efficient Producer Services (EPS). The latter have formed the locus of work undertaken at Cranfield University in this field. A definition of these is provided by Meijkamp (2000):

"Eco-efficient services are all kinds of commercial market offers aiming at fulfilling customer needs by selling the utilisation of a product (system) instead of providing just the product. Eco-efficient services are services, relating to any kind of product, in which some of the property rights are kept by the producer."

This is a specific type of service orientated product concept which on application in commercial environments it is hoped will lead to the development of new service orientated products which satisfy demand at increased resource efficiencies – factor four and above. Importantly, the EPS concept has been designed not only to improve environmental performance but also commercial. In this respect, similar to new service orientated products arising in markets, it is thought that the EPS concept will allow firms to secure additional value. Thus it may be observed that the EPS concept is designed to achieve a 'win-win' and has been developed for the specific purpose of realising the potential that the shift from products to services may provide.

2.2 Development of New Product Service Systems

Whilst a number of manufacturers may have acquired or built organisational competencies to develop and deliver services, application of EPS will require further development of these, to include those which enable for example, life cycle analysis. Thus new product development methodologies have been created to assist in this matter. A number of product development methodologies have resulted from research projects, e.g. Designing Eco-Efficient Services, Creating Eco-Efficient Producer Services. These are based on *innovation*.

Whilst innovation is a generic transformative process, authors have been seeking to identify the specific attributes of an Eco-efficient service innovation process. This work has been based on knowledge of the EPS concept and empirical findings from case studies of instances where manufacturing firms have integrated services into their product portfolio. The notion that it is possible to progress through an evolutionary hierarchy of innovation types is prolific in literatures (Brezet, 1997).

The scheme detailed below is known as the Rathntheau Hierarchy, several types of innovations are defined in this:

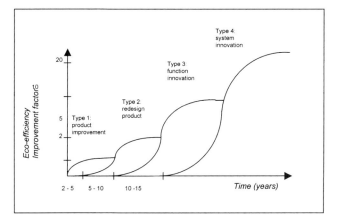

Figure 1 The Rathntheau Hierarchy (Brezet, 1997)

Type One: Product Improvement
Existing products are being adjusted and improved from the perspectives of pollution prevention and environmental care. The product itself, and the production techniques, are generally unchanged. Mainly the adjustments will be, for example, organisation of take back systems for tyres, changing raw materials, changing the type of coolant used, or adding on technology such as catalytic converters.

Type Two: Product Redesign
The product concept will stay the same, but parts of the product are developed further or replaced by others. Products redesign aims at: increased reuse of spare parts and raw materials, or minimising the energy use at several stages in the product life cycle.

Type Three: Product Concept Innovation
Here the way the function of the product is fulfilled changes. A change from paper based information exchange to E-mail, or from private cars to "call a car" systems, are examples of changes in the way a product function is fulfilled.

Type Four: System Innovation
Here new products and services arise, requiring changes in the related infrastructure and organisations. A change over in agriculture to industry based food production, or changes in organisation, transportation and labour based information technology, belong to this type of innovation.

Central to this model is the notion that progression through a number of types of innovation leads to certain improvements in eco-efficiency. Similarly, Vermeulen and Weterings (1996) articulate potential environmental improvements that can be derived from the model:

Phase 1 – Product improvements (up to factor 2 improvements)

Phase 2 – Improvements in existing product concepts (up to factor 5 improvements)

Phase 3 – Innovations in functional fulfilment (up to factor 10 improvements)

Phase 4 – System innovations (up to factor 20 improvements)

Within this model the development of Eco-service products was understood to arise in phases 3 and 4.

2.3 Marketing and Consumer Acceptance of New Product Service Systems

A third stream of research has focused upon consumer acceptance of eco-efficient services orientated products (Cooper and Evans, 2000; Ehrenfeld and Brezet, 2001; Schrader, 1996; Schrader, 1999). In most cases, despite the increasing use of services to satisfy demand, it has been found that these new products require a change in commercial and social habits. As these often entail a change in ownership rights and in particular retention of these by the producer, Schrader (1999) identifies ownership and moreover, consumption without ownership as a critical issue. The table below details opportunities for consumption without ownership:

Table 2 Consumption Without Ownership (Schrader, 1999)

Obstacles to consumption without ownership	Chances for consumption without ownership
Self enhancement by	Self enhancement by
Conspicuous consumption and symbolic self completion Control over material goods	Special experiences, abilities and opinions Ecological consumer behaviour
Orientation by owned goods	Orientation by experiences, abilities or opinions
Freedom	Freedom
By property rights From market transactions	From property duties
Security by ownership	Security by modern economic, social and legal system
	Social needs of single households

Thus it may be observed that a growing literature has been developed in response to the above. Central to much of this, is the notion that the expansion of the service sector in western post industrial economies provides opportunities to improve the environmental performance of products. Research to date has been largely exploratory and inductive in nature. Perhaps as a function of this and relative immaturity, convergence may be found at the highest level and in the broadest sense.

3 REALISING THE POTENTIAL OF THE SHIFT FROM PRODUCTS TO SERVICES IN THE UK

In its recently published consultation paper 'Changing Patterns' the UK Government set out its policy package to stimulate the development of sustainable patterns of production and consumption. This comprises *inter alia* a number of key policy priorities which include, the need to de-couple economic growth from environmental degradation, to work with the grain of markets and identify market failures and to take an holistic approach that considers whole life cycles of products and services, intervening to deal with problems as early as practicable in the resource/ waste flow.

Research undertaken at Cranfield University aims to assess the potential of services and in particular, new product service systems within this policy context. Work has converged around the following research questions:

1. What opportunities does the expansion of the UK services sector and emergence of new service orientated product offerings provide within this policy context?

2. How might these opportunities be realised?

3.1 A UK Based Research Agenda
Research gave rise to a number of key issues. Firstly, there is a need to undertake UK focused research. This would provide insights which are firmly rooted in this context and a robust body of knowledge upon which new concepts and policies can be developed to realise the potential of the expansion of the service sector to improve eco-efficiencies. It is suggested that the following theme should form the loci of research:

3.1.1 Changes in Products
Similar to work undertaken on mainland Europe, product focused research is needed to understand the qualitative aspects of the expansion of the service sector. This would focus on analysis of new service orientated product service systems and particularly instances where material artefacts have become subordinate to services in the product service mix. This would show instances where consumption without ownership is beginning to take root and the basis for opportunities to improve resource productivity.

3.1.2 Changes in Production Practices
Within literatures concerned with services and the environment, there is little emphasis given to wider organisational changes which go beyond those associated with product development but have arisen as a result of integration of services within manufacturers product portfolios. Notions of cognitive bias within the research community may be relevant here, as there is little discussion of notions of capability or competence. It is suggested here that the resource based theory of the firm (Prahalad and Hamel, 1990) may afford particularly useful insights in interpreting such organisational changes. It is hoped that these would form a robust body of knowledge which could be used to identify how new competencies could be developed in order to facilitate development of SPSS and EPS.

3.1.3 Changes in Consumption Practices

Finally, there is a need to identify and understand changes in UK based consumption practices which have arisen as a result of the expansion of the service sector. Similar to work completed on mainland Europe this should focus on instances where consumers are consuming new service orientated products and thus map the opportunities for consumption without ownership.

Also research showed that whilst a number of effective collaborations with industry were developed, research has generally lead to the creation of knowledge within academic and policy making circles. Thus the concepts arising from this work are largely exogenous in nature – developed in the external environment of firms, and subject to associated problems. This appears to have been somewhat overlooked in literature and it is suggested here that this is a fundamental oversight as for the EPS concept to be successful in the UK and improve eco-efficiencies, it must be transferred to and applied within manufacturing companies and, the *transferability* of the concept will to a great extent determine this.

3.2 Transfer and Application

A dedicated stream of research at Cranfield University has sought to explore the transfer and application of the EPS concept. Building on theoretical insights from generic literatures concerned with technology transfer, case study research showed that the factors associated with receptivity largely determined the degree to which the EPS concept could be transferred from academic and policy making circles to UK manufacturing companies. Receptivity is understood here as:

"The extent to which there exists not only a willingness (or disposition) but also an ability (or capability) in different constituencies (individuals, communities, organisations, agencies, etc.) to accept, absorb and utilise a technology"
(Seaton et al, 1998)

Certain attributes of receptivity were found in both the external and internal environments of firms, *inter alia,* these included:

3.2.1 Strategic Orientation of the Firm

Research showed that receptivity arose when the application of the EPS concept contributed toward activities to achieve a firms strategic intent. In this regard, congruence with a number of specific elements of strategy was required, these manifest as the strategic orientation of the firm and arise in response to certain external pressures:

- Changes in market conditions
- Legislation
- Environmental performance

Research showed that strategic orientations were key attributes of receptivity.

3.2.1.1 Changes in Market Conditions

Firms that are receptive to the EPS concept focused on differentiation not cost reduction. In this respect, they intended to achieve this aim through not only technological improvements of their material products but also (of great importance) additional services - they were hoping to achieved greater differentiation through these. The cornerstones of this strategy were:

innovation in both product and service design; and, innovation in the processes through which they approached their market. Representatives from these companies understood that EPS concept offered considerable utility in these respects, e.g. adding economic value through service provision. Thus receptivity arose because the Eco-service concept was congruent with their strategic orientation to changes in market conditions.

3.2.1.2 Legislation
Secondly, a firms orientation to changes in legislation was identified as an attribute of receptivity. Receptivity not only arose when firms were required to comply with 'end of life' legislation but also, when the EPS concept was congruent with a firm's orientation to legislative compliance. Receptive firms argued that changes in legislation created opportunities to expand their business activities. They were pro-active in this respect and viewed changes in legislation as opportunities to expand their business activities and secure competitive advantage. For these firms the EPS concept provided an opportunity to achieve this in response to 'end of life' legislation – the WEEE directive.

3.2.1.3 Corporate Commitment to Improving Environmental Performance
Thirdly, firms that were receptive to the EPS concept had a corporate commitment to improving environmental performance. Also, they argued that this approach to environmental issues created opportunities to expand their businesses and to secure competitive advantage. The EPS concept has been developed as a 'win-win' concept – to improve the environmental *and* economic performance of firms. The coalescence of these objectives matched the orientation to improving environmental performance of these receptive firms. It enabled these firms to meet their strategic intent in this regard.

3.2.2 Organisational Capability
It was found that receptivity arose when their existed those organisational capabilities required for the application and acquisition of the EPS concept. Consistent with the findings of Leonard Barton (1992), a high degree of congruence was required.

3.2.2.1 Application
Receptivity arose when manufacturing firms were already in possession of the necessary capabilities to develop and deliver additional services. In these instances, many of the capabilities necessary to apply the EPS concept had already been built by the firm. These were understood as those required to provide services. Application of the EPS concept required the development of a few additional competencies such as the ability to perform life cycle analysis which, could be acquired through incremental steps.

3.2.2.2 Acquisition
However, the notion that receptivity also arose when there existed the capabilities necessary to acquire and assimilate exogenous technologies must not be overlooked. As discussed above, the EPS concept is in the external environment of firms and therefore receptivity is partly determined by firms' ability to acquire exogenous technologies from this. For example, research showed that receptivity arose in firms which were actively engaged in scanning activities to identify: new external pressures which provided commercial opportunities; new technologies from which they could secure commercial advantage as a result of these. Indeed, receptivity arose in firms which had a strong history and proven track record of successfully acquiring and applying exogenous technologies and management concepts.

4 CONCLUSIONS

It is widely accepted that there exists a link between humankinds economic activities and environmental degradation. Thus any structural change in economic systems is of great interest to those promoting sustainability. In this regard, research shows that expansion of the service sector in western post industrial economies has given rise to a number of desirable changes in patterns of production and consumption and moreover, opportunities to improve eco-efficiencies by at least factor four. A number of service orientated concepts have been developed in policy and academic circles for the specific purpose of realising the opportunities that this structural change may provide and include, *inter alia*, Sustainable Product Service Systems and Eco-efficient Producer Services.

However, little research has been undertaken in the UK and there is a lack of context specific findings. In order to understand and realise the opportunities that the expansion of the UK service sector may provide to improve resource efficiencies this will need to be addressed. This paper has suggested that a detailed understanding of the various aspects of this structural change needs to be developed and that this would require further research to understand associated change in: products, production practices, consumption practices.

Also and importantly, the paper has stressed the need to recognise that research to date has lead to the creation of knowledge in the external environment of firms and therefore the EPS and SPSS concepts are exogenous in nature. In order for these to be successful they must be transferred to and applied within manufacturing firms. This paper reported the findings of research undertaken to understand identify the attributes of receptivity which relate to the EPS concept.

In conclusion, whilst the expansion of the UK service sector may offer great potential to improve resource productivity, in light of current research form a variety of field, further work is required to identify opportunities and how these might be realised.

REFERENCES

1 **Brezet, J., Bijma., A. and Silvester, S**. (2000) Innovative electronics as an opportunity for eco-efficient services. Electronics goes green 2000+, Berlin, Germany, 859 – 866.

2 **Brezet, J**. (1997). Dynamics of Eco-design practice. UNEP IE: Industry and Environment, 20 (1-2), 21 – 24.

3 **Cooper, T., Evans, S**. (2000). Products to services. Friends of the Earth.

4 **Department for Environment Food and Rural Affairs.** Changing Patterns: UK Government Framework for Sustainable Consumption and Production. HMSO, London, UK

5 **Giarini, O**. (1987) The emerging service economy. Pergamon Press, Oxford, UK

6 **Leonard Barton, D**. (1992). Core capabilities and core rigidities: A paradox in managing new product development. Strategic Management Journal. 13 (Special Edition).

7   **Manzini, E.** (1998) Pioneer industries on sustainable services. Introductory notes. A paper presented to workshop entitled "Industry and Sustainability: Pioneer Industries on Sustainable Services" at the INES conference: Challenges of Sustainable Development.

8   **Meijkamp, R.** (2000). Changing consumer behaviour through 'eco-efficient services' – an empirical study on car sharing in the Netherlands. PhD thesis, Delft University of Technology, Delft, Netherlands

9   **Ottman, J. A.** (1998). The development of sustainable products and services. Corporate Environmental Strategy, 5 (5), 81 – 89.

10 **Prahalad., C. and Hamel, G.** (1990) The core competence of the organisation. Harvard Business Review, 79 - 91

11 **Schrader, U.** (1996). Consumption without ownership – a realistic way towards a more sustainable consumption? 5th International Research Conference of the Greening of Industry Network, Heidelberg, Germany, 1 – 20.

12 **Schrader, U.** (1999) Consumer acceptance of eco-efficient services – a German Perspective. Greener Management International, 25 (Spring), 105 – 121.

13 **Seaton, R., Jeffrey, P., Stephenson, T. and Parsons, S.** (1998) From Marketing to Receptivity: Structuring Community Involvement in Integrated Water Management." WATERTECH, Brisbane, Australia.

14 **Trott, P.** (1993) Inward transfer as an interactive process: A case study of ICI. PhD Thesis, International Eco-Technology Research Centre, Cranfield University, Cranfield, UK.

15 **Vermeulen, W. and Weterings, R.** (1996). Van afvalzorg naar milieu-innovatie van produkten; Een visie op producentenverantwoordelijkheid. Rathenthau Instituut, Den Haag

16 **Weizacker, E., Lovins, A.B., and Lovins., L. H.** (1997). Factor Four – Doubling Wealth, Halving Resource Use - the new report to the Club of Rome, Earthscan , London, UK.

17 **White, A. L.., Stoughton, M. and Feng, L.** (1999). "Servicizing: the quiet transition to extended product responsibility." Tellus Institute, Boston.

Extending product life – the forgotten challenge

R DODDS
Department of Engineering, University of Liverpool, UK

ABSTRACT

The extension of the life of products is one of the most obvious routes to reduce the environmental load. It was recognised explicitly in the Limits to Growth publication of 1972. This route is not pursued with the ruthlessness and enthusiasm that other approaches, such as recycling, have inspired. This paper considers why this should be the case and explores the possible circumstances that could place this approach nearer the top of the sustainability agenda.

1. INTRODUCTION

In the chapter 'The State of Global Equilibrium' of the seminal work 'The Limits to Growth' for the Club of Rome in 1972 [1] there is a list described as…" a few obvious examples of the kinds of practical discoveries that would enhance the workings of a steady-state society" :

- New methods of waste collection
- More efficient techniques of recycling
- Better product design to increase product lifetime and promote easy repair
- Harnessing on incident solar energy
- Methods of natural pest control
- Medical advances to decrease the death rate
- Contraceptive advances to facilitate equilibrium of birth and death rates

Six of the seven have received considerable attention and success, but the world has apparently been particularly silent on the third. This paper questions why this is the case and what circumstances would need to exist to generate activity in this area.

It is useful to consider how the business, political and social environments have changed significantly in recent years. For example :

Consumers are becoming more sceptical of multinationals, governments and 'environmentalists', all being perceived as having vested interests, but at the same time they are becoming more aware of the relationship between their purchasing choices, their way of using the product and the resulting impacts both on the environment (for example, the choice of automobile, the type of fuel and fuel usage) and on wider society (for example the emergence of growing 'ethical' consumption niches in particular product areas such as organic foods and Body Shop). At the same time 'branding' is becoming more deep-rooted with global communications and advertising strengthening prestige brands and so-called designer products which becoming increasingly important for the young.

Businesses are realising that in many cases profit will come increasingly from leasing/service elements and less from the initial product sale.

Governments, particularly in Europe, are appreciating the adverse impact of high product turnover on the environment and are intervening with new legislation which puts the final and significant disposal responsibilities back to the manufacturer or retailer for refrigerators automobiles and wide classes of electronic and electrical goods. Indeed, legislation tends to being forced through before facilities and business processes are available.

2. THE BENEFITS OF AN INCREASE IN PRODUCT LIFE

From a 'usage of materials' perspective, an increase in product life is indeed attractive. An increase of product life by 50% will result in a reduction of materials usage of 33%. It does not seem unreasonable to extend the life of automobiles from say 12 years to 18 years, the life of washing machines from say 6 years to 9 years, the life of mobile phones from say 3 years to 4.5 years – the aggregated cumulative impact on materials usings would indeed be significant.

We have come a long way since the publication of Vance Packard's 'The Wastemakers' [2] brought public attention to the fact that products were being designed for a limited life – planned obsolescence - to ensure repeat sales.

Although manufacturers now have to compete on product performance and operational costs , industry have continued their quest to sell more products. Manufacturers still try and persuade consumers to purchase a new product before their current product has worn out. Repair costs often appear high to the consumer, encouraging replacement . The extension of product life is certainly not a business driver. Manufacturers deal with environmental concerns by finding more acceptable sources of raw materials and by entering into government-supported schemes to encourage recycling.

Whilst the ideal scenario perhaps is to leave materials in the ground, and to ensure those that must be taken out remain above the ground and serving a useful purpose for a long as possible, before returning as land-fill, the current focus on recycling would suffice if it provides a long-term route towards sustainability.

Let us question current approaches by taking an area where recycling is being taken seriously- End of Life Vehicles - with impending EU legislation to force the responsibility for automobile disposal back on to the manufacturers through a range of measures: the establishment of vehicle de-pollution stations, formal logging of the production, sale and then disposal of each car, enforced materials composition limits to aid recycling and reduce land-fill. The system likely to emerge is that cars will be depolluted and 'shredded' and the various component materials will be re-used. The metal mixtures will be exported to less developed countries. This appears logical at face value, assuming the manufacturers/consumers pick up the extra costs, but the situation will change when receiving countries have no further use for low-grade metals and prefer to use that from their own scrapped cars. At this stage more sophisticated recycling and re-use methods will be needed with significant cost impacts. And significant uses or alternative disposal methods may not be found for the increasing levels of non-metallic materials (foam/polymer) mixes currently destined for land-fill.The legislated solutions are therefore perhaps only temporary. The choice between recycling, reuse and extension of life deserves more consideration[3].

3. WHY IS THIS APPROACH NOT BEING TAKEN

There are several reasons why extension of product life is not being embraced by industry :

3.1. Business Success is measured by Volume Growth.
One can speculate that the main reason for the lack of attention to the extension to product life is that this approach runs counter to the fundamental business paradigm of success being measured by growth. The enormous pressure for business success through growth are set by the financial markets with short-term perspectives.

This growth is generally achieved by economies of scale resulting in products being more affordable to a greater number of consumers, and then from generating a continuous stream of gradually improving products that the consumer wishes to purchase, discarding the previous version. Business appears comfortable with tackling the increasing number of redundant products by improving the capability for recycling materials.

Of course, the impact of selling 33% less product would have a drastic effect on the number of people involved in manufacturing and selling but it is the reduction in revenue and profit that prevents companies from moving in this direction, rather than their social conscience.

3.2. Consumers' rising expectations
Consumers, encouraged by persuasive sophisticated advertising campaigns and new communication routes, now expect access to a wide range latest consumer products and prestigious brands. People no longer expect to start a home with second-hand furniture and hand-me downs. Plentiful credit, economies of scale, slick distribution and retail systems have made affordable a much wider range of goods, which were previously 'out-of-reach' of the average consumer. Fashions, automobiles, electronic goods are all examples where the time delay has reduced dramatically between these goods, or their novel functionality, being affordable only by the 'rich' to coming within reach of the average-income consumer.

3.3. Step-changes in technology make wholesale change-over beneficial.
There are two types of circumstance where significant technology change can result in a situation where wholesale replacement of a class of products can be beneficial.

Firstly, those situations where the functionality is rapidly changing which normally occurs in the early years of a product type. Increasing market penetration and improved economies of scale in manufacturing and distribution help to make these new products and product upgrading affordable. Currently camcorders and digital cameras fall into this category. Mobile phones are perhaps moving out of this category, and in general hi-fi systems, TVs, video recorders and most white goods have moved out. The technology changes in this latter group are generally not sufficiently distinct to justify an upgrade unless the existing model has broken down or become unreliable.

Secondly, those products where design changes have resulted in a significant change in the environmental footprint of a product. The environmental footprint is often dominated by the 'use' stage, whether an automobile or a washing machine. For example, significant reductions in power or fuel usage or fuel type due to improved design could justify a change. The relatively recent star-rating of the energy-efficiency of refrigerators has made consumers aware of this aspect and in spite of the relatively low impact of the power used by a refrigerator on household expenditure budget, the inability to track the power usings through the household electricity meter and the lack of a favourable purchase price, the swing to low-power fridges has been significant and is clearly beneficial to sustainable development. Some form of government intervention is normally needed to carry this type of change through.

4. POTENTIAL BUSINESS AND CONSUMER RESPONSES

The prospect of businesses accepting longer product life as a key objective seems bleak since it challenges the current business paradigm of growth.

However, businesses respond to consumers. If consumers accept holding products for longer periods, then solutions are more likely to emerge. Let us consider some examples outside consumer products where long product life is not viewed negatively:

One doesn't throw away old masters or rare antiques. These are prestigious to own, scarce with no further supply, and therefore go up in value. In fact 'the older the better'. The brand here is the artist or the date of origin. Additional supply is zero.

One doesn't knock down attractive old buildings considered to be part of our heritage. There is a prestige value in living in a well-kept building with its own history. Those that have deteriorated or become unaffordable to maintain have some level of protection or support from to government intervention.

Luxury automobiles such as Rolls-Royce and Bentley have enormous prestige value since they can be afforded by only a small proportion of the population. They are built to last. The owners can afford any necessary repairs. Running costs don't matter. Since they are scarce and coveted, they hold their value well relative to other vehicles.

The common element in the above examples is the pride of the owner in continuing to own these products and the relationship developed between the owner and his paintings, listed building or luxury car.

In summary at this stage, we could suggest necessary but not sufficient conditions for consumers to be happy to hold products for longer periods are that:

- Functionality and efficiency are retained and not drastically outperformed by new models
- There is no negative status of being seen with an older product
- The consumer has developed an interactive relationship with the product.

There are a number of approaches that could be taken to meet these conditions. For example by:

- the product being a brand valued by that consumer or designed by a prestigious figure
- the product being personalised through the design of the product
- the product being personalised through the interactive operation of the product
- the product being updated operationally 'within its body' by the manufacturer

If these consumer demands became apparent, business would surely block them until they could see a way of maintaining revenue and profit growth. The way this might be achieved is by the manufacturer and retailer replacing their revenue from the sale of new products with revenue from servicing the product.

Successful examples are unusual.

The obvious route is maintenance costs are a significant item for the current car owner and a significant revenue stream for the industry as the technical complexity has increased so that replacement of faulty modules rather than repair becomes the norm. The aircraft industry has successfully introduced 'power by the hour' contracts where the user can predict the operational and maintenance costs and avoid the costs associated with unpredictable breakdowns.

The computer printer market is an interesting example where the main revenue is now from the sale of printer cartridges, not the printer. Some estimate that the cartridge business accounts approaches 90% of the printer market. However to achieve this the printer cartridge margins are so high that consumer resentment emerges, as well as less-expensive printer cartridge refilling shops.

In asking whether these principles can be extended to other products we should note that computer printers and mobile phones for the consumer are relatively new and business relationships are fluid. In more traditional products such as automobiles and white goods, where such approaches might be beneficial, factories and business relationships have become entrenched and there will be a massive reluctance to change.

However, there have been changes in established industries. Most large automobile manufacturers have in recent years embraced and upgraded the second-hand car market by introducing 'approved used car schemes' with various financing options. There are sustainability benefits, whether intentional or not, in that the period of time for which the car is maintained in good condition is extended, and therefore its life is likely to be extended, although not with the original owner.

However, none of these potential business responses would encourage consumers to hold product for longer. Consumers are driven by cost far more than concerns on sustainability. If consumers had on-line information about the performance levels of their products and knew the effect of their user habits on running costs and product life, interaction with the product would become financially beneficial to the consumer. How would manufacturers view this?

5 THE OPPORTUNITY FOR NEW BUSINESS MODELS BASED ON PERFORMANCE

Business models have changed from sale, sale and warranty, to sale with extended warranty plus perhaps through breakdown insurance policy. The manufacturer/retailer is therefore taking on a higher proportion of the risk and passing the cost onto the consumer. The event being covered is breakdown and not operational performance.

The manufacturer would argue that making a financial commitment to a performance level depends on how the product is used – fuel consumption and tyre wear depends on how the car is driven; number of wash loads, water and power usings of a washing machine depend on weight of load, wash cycle selected, type of soil in the clothes and incoming water temperature.

However, technology is now emerging which could enable both the manufacturer and user to observe and record current and cumulative performance levels and get an estimate on remaining product life. Internet and wireless communications will enable manufacturers or an intermediary to gain access to this information and also to update the software within the product to increase its capability. Physical component replacement and upgrading is of course an option.

Let us take some examples:

In an automobile, this could include fuel usage levels (mpg) and tyre life (miles driven). Since these vary with driving habits, these driving habits would need to be monitored for each car and costed accordingly.

In a washing machine, this could include the number of washes, the water usage, the power usage. Since these vary with the type of soil, the wash load size, type and condition, these would need to be monitored for each machine.

In a computer printer, this could include the number of sheets printed before a cartridge is needed. Since this will vary with the colours used, font size and quality demanded, then these would need to be monitored for each machine.

In order to change the attitude of the consumer and manufacturer to product life and encourage longer ownership, we need to measure and make available to the consumer, not what has been used, but what is left to be used, and how the consumer-use habits and quality of design affect 'what is left' .

If products can be identified with these interactive capabilities which appeal to a sufficient number of consumers, then manufacturers may respond with favourable and imaginative combinations of selling price/leasing offers involving guaranteed performance levels of the product; performance levels which relate to a manufacturers commitment on the operational costs incurred by the consumer and the expected product life.

Would manufacturers respond to this challenge ? Nowhere is this balance between functionality, user habits and cost more explicit to the consumers than in the current mobile phone market where a whole range of charging mechanisms through contracts are being

D004/023/2004

offered in an extremely competitive market. The mobile phone user is very aware of the relationship between the way the phone is used – to whom, where, for how long, on what receiving phone – and the cost of use. The itemised bill received each month brings this all vividly home. Heavy users get the next phone free (and hand in the old model to be professionally disposed of) since the supplier's revenue stream and profit is mainly from the usage charges. But surely this was initiated by competition between the intermediaries rather than the manufacturers, and so perhaps the utilities and fuel providers would need to be potential partners in any such initiative.

6 CONCLUSIONS

Technology is emerging that will permit personalised and updatable product systems and meaningful usage and user-habits monitoring. This information could be used to put pressure back onto the manufacturer to make a contractual commitment on operational efficiencies and product life. Both of these aspects hit consumers in the pocket through running costs and the impact of an often unexpected replacement cost. Mechanisms to encourage consumer interaction through their products with manufacturers and intermediaries need to be explored in order to move towards truly useful cost-related information which will bring more attention to the benefits of extended product life.

REFERENCES

1. Meadows D.H. et al (1972). 'Limits to Growth', a Report for the Club of Rome. Pan Books. ISBN 0 330 2416 99.

2. Packard V (1960). The Wastemakers. Van Rees Press.

3. Stahel W.R. (1994) The Utilisation-Focused Service Economy: Resource Efficiency and Product-Life Extension. The Greening of Industrial Ecosystems, National Academy Press.

Sustainable Design Education

Factor 10 – the changing landscape of learning

E DEWBERRY
International Ecotechnology Research Centre, School of Manufacturing, Cranfield University, Bedford, UK

ABSTRACT

A Factor 10 increase in the productivity of natural resource requires transformations in the way we all learn, work and play. Its central goal highlights the importance of individual accountability - the cause and effect of our day-to-day decisions about what material and energy we choose to buy, use and waste.

These changing perspectives on utility are explored in this paper through a learning process aimed at encouraging the generation of more sustainable products, processes, messages and systems. The components of this pedagogical approach are described alongside learning outputs that aim to move us closer toward Factor 10 living.

1. BEYOND EFFICIENCY

Sustainable development is about all aspects of ecological, human and economic wellbeing and the aim to balance these at both local and global scales. Our current development paradigm focuses on a quantitative measurement of success, made explicit in the value of the monetary wealth of economies and individuals. Less quantifiable goods such as love, friendship and community that add greatly to overall notions of our wellbeing are, for the most part, ignored and at best, undervalued - *'...the unhealthiness of our world today is in direct proportion to our inability to see it as a whole'* (1).

Thus the process of sustainable development requires us all to think differently about what we do, how we do it and why we do what we do. While this may sound a straight forward enough task there's plenty of evidence out there - climate change, loss of biodiversity and increasing social dislocation - to suggest otherwise. The inertia to change tact – particularly from key political, business, economic and social agencies - is possibly a predicable reaction given that problems of unsustainable development require inherently complex responses that look to radical changes within and across the existing system. Therefore the challenges involved in implementing a more sustainable development agenda are directly linked to the way in which people perceive the risk of such change and how the outcomes of such change are explained. To date much of this agenda has been translated in legislative and policy based language and has generally limited actions to those associated with incremental efficiency gains in product and process improvement and end-of-pipe clean up. However, to achieve a Factor 10 improvement in resource use as part of a sustainable development process, a more holistic view of both supply (manufacturers' responses to sustainability) and demand (consumers' reactions to sustainability), and the links between the two is required. A move *beyond* efficiency to also address sufficiency is critical. Factor 10 involves a radical transition in the hearts and minds of people – particularly those living in the industrialized nations. This scale of improvement equates to a 90% efficiency gain, in other words achieving the same results on 10% of current resource use. Many commentators suggest that it is this level of improvement in resource use that is required over the next fifty years (from 1995) to begin to turn around unsustainable development (2, 3). Frederich Schmidt-Bleek, the founding President of the Factor 10 Club, emphasises this point further, "*Despite the prevailing uncertainties I remain convinced that if the process of dematerialisation does not begin soon, both the social fabric or our societies and the global ecosystem are seriously at risk in the medium term.*" (4)

The scale of response required for sustainable development points toward solutions embracing individual, as well as group, motivation and accountability. Factor 10 sustainable change necessitates people to think and act differently which in turn suggests a need for people to be made more aware of the issues and given the skills with which to deal with such change. It seems likely therefore that education could be transformative in encouraging different perspectives across the whole lifecycle of products and systems (supply and demand), and that this goal requires a shift away from current educational practice. It involves a journey from a teaching-centred, problem solving and 'controlled' pedagogical approach toward a learning focused, problem reframing, responsive and dynamic approach to education; one that embraces joined-up thinking and promotes a range of alternative responses (5). *In thinking about the kinds of knowledge and the kind of research that we will need to build a sustainable society, a distinction needs to be made between intelligence and cleverness. True intelligence is long range and aims towards wholeness. Cleverness is mostly short range and tends to break reality into bits and pieces. The goal of education should be to connect intelligence with an emphasis on whole systems and the long range cleverness, which involves being smart about detail* (6). This paper describes the embryonic development of an educational approach addressing design for sustainability from a perspective of 'the whole'.

2 SUSTAINABLE DEVELOPMENT EDUCATION

A number of reports highlight the key role of education in enabling transitions toward sustainable development. Agenda 21 (7) states '*education is critical for promoting sustainable development and improving the capacity of people to address environment and development issues.*' In 1990 twenty-two university leaders convened in Talloires, France to voice their concerns about unsustainable development and to establish the need for action from the higher education community: '*Universities educate most of the people who develop and manage society's institutions. For this reason, universities bear profound responsibilities to increase the awareness, knowledge, technologies and tools to create an environmentally sustainable future.*' (8). The Talloires Declaration has since been signed by more than 265 university presidents in over 40 countries across five continents. In the UK The Toyne Report, *Environmental*

 D004/014/2004

Responsibility (9) stressed the need for Further and Higher Education (FHE) institutions to not only introduce sustainability curricula into all courses but also to make the institutions themselves, their campuses and their purchasing policies, more sustainable. A subsequent UK Government publication, *Sustainable Development Education: Design Specification* (10) identified three main areas of focus in the development and implementation of sustainable development education strategies: sustainability concepts; sustainability solutions; and effective teaching.

Sustainable development requires an educational approach which prioritises connected, non-mechanistic thinking and where practice and theory are viewed as mutually supportive actions. According to Ali Khan (11), the pedagogic approach for sustainability necessitates a shift away from a teaching to a learning paradigm that, '*emphasises independence of mind and the ability to make sense of, rather than reproduce, information*'. Sterling (5) expands this view in addressing the different approaches and values of transmissive (transfer of information to learner) and transformative (learner constructing and owning meaning) methodologies '*that go beyond teaching method to also reflect philosophy and purpose of education.*'

Table 1. The differences between transmissive and transformative education (5)

transmissive	transformative
instructive	constructive
instrumental	intrinsic
training	education
teaching	learning
communication of message	construction of meaning
information focus	appropriate knowledge
central control	local ownership
product oriented	process oriented
problem solving	problem reframing
linear	iterative and responsive
facts and skills	concepts and capacity building

Sterling argues that 'education for change' is complex and involves ambiguity and uncertainty and as such alternative learning strategies are required to make sense of this new learning landscape. This new context for learning embraces trans-disciplinary, participative, creative, constructive and responsive methods that allow for (and respect) new perspectives and understanding and the continual reflection necessary for problem reframing and capacity building. It is about the engagement of the individual and the whole learning institution in meeting the challenges and opportunities for change.

Interestingly, best practice design education shares some key characteristics with a transformative education approach. It involves creative, solutions-focused learning; self-directed team work; learning by doing (commonly 'live' projects); iterative refinement and reflection; and, developing research and interpretation skills drawing on multiple sources. It therefore provides an interesting platform on which to apply a sustainable education philosophy. It is this combination of design and sustainability pedagogies that are explored in the following section.

3 EXPLORING A SUSTAINABILITY AND DESIGN CURRICULUM

The basic remit of design for sustainability is to generate as much utility and enjoyment as possible out of the smallest possible quantity of natural resource over the longest possible period of time. To achieve such a tall order design can no longer just be about the material, the production, the function, the aesthetic - that which the majority of UK design education still focuses on. Designers and those involved in creative and innovative industries need to address whole lifecycle concerns if increased utility per unit resource over an extended period of time is to be delivered. These lifecycle issues include the origin of resource, the nature of energy used and pollutants produced in extraction and production, the environmental and social impacts of resource, product and service travel, consumption culture and methods and adaptation of design outputs in use, the cultural impetus to dispose, to seek the new, and the collation of resource and the energies expended in dealing with reuse, regeneration and disposal. These issues not only embody supply issues (those of the traditional manufacturing and commerce remit) but also those of demand (consumer behaviour, cultural expectations and trends), those which have not truly been integrated as yet in design for sustainability but which need to be to enable Factor 10 utility increases to occur.

When we begin to challenge the current paradigm of development (as sustainability requires us to do) we will desperately need people who have the ability to think creatively and laterally and draw on disparate areas of knowledge and to vision new, more sustainable futures. These people are the designers of the future. It is the responsibility of current design education to begin to embrace these ideas and develop the pivotal role of design for sustainability in the visioning of sustainable scenarios for this century and beyond.

3.1 Integrating Sustainability in UK Design Education
The problem is how to integrate such diverse issues within existing design education (that supplies designers to the existing requirements of the market which are not working to a Factor 10 agenda). If we take Sterling's view of transformative education, the solution resides in the ability to encourage a participative and reflexive learning path; to look at a diversity of knowledge required in a lifecycle approach to design from many perspectives; to make new connections and develop new links between previously disparate sets of knowledge.

In the UK there are few universities that address these core components of sustainable development education and within UK design curricula, even fewer departments fully addressing the potential of design for sustainability (12). This is because it is hard to do and deeply challenges the cultural mindset of what design is and what designers do. However the pressure to integrate sustainability is growing and initiatives like the Sustainable Design Award for As & A2 Level students run by the Intermediate Technology Development Group (ITDG) and facilitated through Loughborough University, explore a wide range of design and sustainability ideas. As a result scholars progressing onto undergraduate design programmes will have already experienced the rationale and expectations linking design and sustainable development.

D004/014/2004

demi (design for the environment multimedia implementation) was a recent (1998-2001) TLTP sponsored project that aimed to integrate sustainability within undergraduate design curricula. Details of the development of the *demi* web-resource (www.demi.org.uk) are described in Fletcher and Dewberry's paper in the International Journal of Sustainability in Higher Education (13). In summary the project mapped out key areas of focus for a design curricula around the six sustainability principles of: efficiency; appropriateness; sufficiency; equity; systems; and scale. It then cross-referenced within and between these areas (linked to information in the web-resource and also to information and experiences in external organisations and resources) to build learner knowledge which was then connected to real projects and integrated to opportunities for future scenarios and projects. As with all prototypes *demi* requires updating and re-visioning, but within it reside some key building blocks for design for sustainability education.

3.2 The Cranfield Course

The remainder of this paper however focuses on a post-graduate perspective on design for sustainability. At Cranfield University in the UK a relatively new course has been introduced within an existing Manufacturing Masters programme. The course, MSc Manufacturing: Sustainability and Design, is situated in the International Ecotechnology Research Centre, part of the School of Manufacturing. It began in 2001 and is currently running with its third group of students. The course aim – in its widest context - is to enable a shift in thinking, toward sustainable development goals, within organisations and society at large. The course emphasis on design for sustainability means this goal is communicated through developing learners' abilities to:

· assist organisations in their understanding of sustainability;
· manage projects which aim to implement change;
· develop design for sustainability strategies for organisations;
· assess the environmental and social impacts associated with products, services and systems; and
· vision scenarios and actions that reduce environmental and social impacts.

The course ethos aims to encourage learners to explore sustainability as 'a way of thinking' and to evolve their own interests and responses throughout the course and beyond. The course is building towards a model of transformative learning (it isn't there yet). The confines of the University's modular course system mean that the course must adhere to an established structure of learning which predominantly involves week-long teaching modules – this also offers opportunities to part-time (professional) learners who can select a range of modules to fit in with their own schedules. The onus then on a truly reflexive and integrative form of learning is difficult to achieve given the modular nature of learning. However the essence of a transformative approach is increasingly being addressed through the following: building connections between modules (e.g. the same case studies viewed from the different perspectives of each module; small learning groups enable ownership and participation; integration of local projects and issues and the importance placed on understanding issues of scale; the focus on process and systems thinking; and the drive to encourage each learner (including staff) to reflect on and develop their own perspective in this area. These learning components are further combined throughout the course through the many projects undertaken. These include short projects (in the form of assessment for some modules e.g. mapping food systems on campus and developing alternative processes and products to address the environmental and social impacts), to three month long group and individual thesis projects that usually involve working with external agencies (e.g. addressing corporate sustainability in a global consumer electronics company; exploring social exclusion issues for teenagers in the North East).

Figure 1 is a diagrammatic representation comparing the evolving course pedagogical approach with a traditional model of learning. The left hand side box and line within it represents the traditional learning path, confined within a specific discipline The student travels this path throughout the course (time A to B), absorbing information and linking and reflecting on that information to achieve a higher level of understanding in their chosen area. Information is broken down into smaller bits taught in a direct way that enables students to absorb the information in a continual stream. Little attention is placed on the students' own learning agendas and needs and the essence of the teaching structure is to encourage throughput and output in the quantitative measurement of learning and thus in the number of 'successful' students. This type of learning effectively produces 'clever' people, responding to market conditions which unfortunately do not as yet integrate sustainable development priorities.

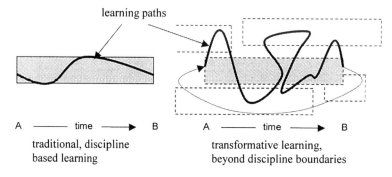

learning paths

| A ——— time ——▶ B | A ——— time ——▶ B |

traditional, discipline
based learning

transformative learning,
beyond discipline boundaries

Figure 1 Different learning paths

The right hand side box represents the move towards a more integrative and reflexive approach to learning within the Cranfield course. The broken line of the box illustrates the acknowledgement that the learner needs to look outside to draw on a range of other knowledges (e.g the other boxes shown). These other knowledges may include formal knowledge (e.g. ecology; complex systems theory; psychology) and more experiential knowledge found in our own personal experiences, in local knowledge, in cultural wisdom. It is the combination and balance of both that helps build a more holistic pedagogy. The line represents the path between these knowledges although still constrained with the time-span of the course (A to B). However what we have found is that past students are happy to come back and share with present students their experiences - and their problems, in terms of projects for the course - and insodoing they themselves reflect on their own learning and development during and beyond the course duration.

The course has an inherent requirement to seek new connections that 'break down' the walls between established disciplines. These connections, framed in a context of sustainability, help

D004/014/2004

formulate questions that encourage alternative responses to current modes of living and working and maximise the potential of design thinking to influence the environmental and social impact of goods and services. Examples of some of the problems addressed by students on the course are explained in more detailed in the next section.

4. LEARNING THROUGH PROJECTS

Cranfield University has a history of translating formal academic knowledge to application based knowledge that can be integrated within organisations. The Sustainability and Design Masters supports this aim by exploring how our learning approach can influence sustainable change – both in organisations and society at large. Indeed, being able to influence people and organisations and vision alternative solutions to their problems is at the crux of implementing Factor 10. The course therefore aims to focus a lot of its time in addressing such issues through the projects its students undertake. The following describes a project undertaken by one of the students from last year's group.

4.1 The investigation of renewable material integration in a product and the development of appropriate communication and training strategies
This project was undertaken by Debra Lilley (14), a student on the second year of the course 2002-2003. It involved investigating the opportunities and concerns for the integration of a new product line that combined renewable and non-renewable materials and where the marketing focus would be driven by the concept of renewables. Although the company had a strong sustainability philosophy it sought an external view on the product launch to address the synergies between product, internal and external markets, and communications between the lifecycle agencies. This project focused on the perceptions of designers and architects ('the clients') and also the perspectives of internal company personnel involved in communicating new messages on renewables. An overview of the project is described in Table 2.

The student carried out an extensive literature search to: build a context for renewable integration; to explore the technological and social issues of renewable use; and to understand internal 'sustainable communication' strategies within the company. This involved the student combining knowledge from a wide variety of disciplines (e.g. ecology; business; systems (design for sustainability); marketing; biology) through secondary research investigation. It also involved a number of conversations with internal personnel to clarify their perceptions on sustainability, communications and training and current market perceptions of renewables (and the proposed product). Issues arising from this phase of the research included:

- The timescale for biodegradation - dependant on size, shape and application of new renewable component;
- Concerns regarding the viability of introducing composting on a large scale;
- The lack of supporting infrastructure;
- The potential increase in release of methane gas which may follow widespread adoption of biodegradable biopolymers being disposed of in landfills at the end of their life;
- The external market reflects that bio-based products are still regarded as relatively expensive in comparison to conventional polymers; and

- The gap between environmental concern and purchasing behaviour.

Table 2. Overview of Research Project (14)

Project Phase	Research Tasks
Background Research	1. Investigation of the development of renewable materials 2. How the drive towards renewable materials has been addressed internally (within the organisation) and externally (wider industry) 3. How renewable materials are communicated/reported upon within the design community.
Action Research	1. Investigation of the perceptions/understanding of architects and designers of renewable resources 2. Investigation of: a) the decision making process b) the drivers for architects and designers to consider biopolymers c) the incentives/benefits they feel this material offers to their end clients.
Synthesis Of Research, Conclusions And Recommendations	1. Conclusions drawn from background research and action research. 2. Recommendations for a marketing strategy for 'the product' as a product and concept 3. Development of a Sustainability training package for salespeople.

In interviews with designers and architects it became apparent to the student that communication and training were important issues for the company to address when launching a 'radical' new product. She had identified from initial research that the company currently focused their communication of sustainability initiatives and their current training programme on a corporate level as opposed to a product level. Their marketing approach also emphasized company level sustainability oriented processes rather than sustainability information on the product itself. This approach was misaligned with that of the market where clients sought product based sustainability information. The in-depth analysis of transcripts and subsequent synthesis of this knowledge through mapping and other techniques (e.g. see Figure 2) produced a comprehensive set of outputs to aid 1. the communication strategies for the introduction of a renewable based product and 2. the sustainability training and education of sales personnel.

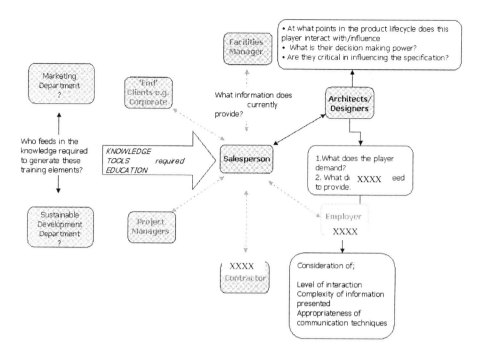

Figure 2: Stakeholder Expectation Analysis; Tracking Flow of Information and Communication Channels (14)

A key outcome of the research findings was the decision by the company to concentrate on the renewable resource benefits of the product, without focusing specifically on biodegradability as an issue. This decision was reached due to concern over the 'lack of completeness' of the lifecycle story, in particular the emphasis on the barriers associated with a lack of supporting infrastructure for the 'end of life' stage.

This project demonstrated a 'problem reframing process'. The research findings challenged the existing perceptions of those managing the product launch in the company, of how a renewable product would be viewed both internally (by sales) and externally (by clients). The student not only answered the brief which focused on the specifics of the renewable product but also refocused the research to answer new questions as they emerged – hence the later emphasis on training and education of staff and the link between this, the company level sustainability philosophy and the need for more detailed product based sustainability information in the marketplace. The student constructed and continually reflected and refined the learning process to produce a comprehensive agenda that effectively addressed the company's needs.

5. A FACTOR 10 FUTURE?

The Cranfield course has some way to go to achieve a truly transformative approach to learning. It c urrently c an only provide a time based learning structure because of the constraints of the University's one year modular masters format. It cannot reside completely in an experiential context where students draw from practice and their and others experiences – but is does offer some of this in the projects the students choose to tackle. It does focus on process rather than product and it does support problem reframing in preference to problem solving. The ethos of the course is one which addresses 'wholes' not 'bits'. Originating from the design perspective of lifecycle thinking, it has developed to appreciate the need for a pedagogical approach embodying participation, ownership and the building of new and alternative connections and combinations of knowledge. In this way we hope the course will be one, among many, that nurture the designers of the future in working towards Factor 10, *'Advanced resource productivity requires integration not reductionism – thinking about the design challenge as a whole, not as a lot of disjointed little pieces'* (15).

Finally, acknowledging the need to 'move minds' and the associated concept of paradigm shift (where Factor 10 requires both cultural and technological movements in perspective) it must be realized that those in education and those involved in any form of learning and reflexive process (everyone!) have the potential and the creativity (and the responsibility) to seek new solutions, grounded in the knowledge that we are aware of what we currently do and where we need to move to. As the physicist David Bohm neatly puts it, *'Thought thinks pollution is a problem "out there" and it must solve it. Now that doesn't make sense because simultaneously thought is creating all of the activities which make the problem in the first place and then creates another set of activities to try to overcome it. Thought doesn't stop doing the things which are making the ecological problem .... Now I think that is really why it is so hard to put a new consciousness into practice, because we are unconsciously in our practice doing the opposite of what we claim we want to do. Therefore the important point is to be aware of what we are actually doing.'* (16).

REFERENCES

1. **Senge P**. (1990) *The Fifth Discipline*, Doubleday Currency, New York.

2. **World Resources Institute** (1997), *Resource Flows: The material Basis of Industrial Economies*, WRI, Washington D.C.

3. **Weaver P. and F. Schmidt-Bleek** (eds) (2000), *Factor 10 – Manifesto For A Sustainable Planet*, Greenleaf, London.

4. **Schmidt-Bleek, F**. (2000) *Factor 10 Manifesto*, http://www.factor10-institute.org/, accessed 6th June, 2004

5. **Sterling, S.** (2001) *Sustainable Education: Re-visioning Learning and Change*, Schumacher Briefing No.6: Schmacher Society, Green Books, Dartington.

6. **Orr, D.** (1994) *Earth in Mind*, Island Press, Washington D.C.

7. **UNCED** (1994), *Agenda 21, Chapter 36 Education, Training and Public Awareness*, United Nations Conference on Environment and Development.

8. **University Leaders for a Sustainable Future,** (1990). *The Talloires Declaration*, ULSF, Washington.

9. **Department for Education** (1993), *Environmental Responsibility: An Agenda for Further and Higher Education*, HMSO, London.

10. **DETR and Forum for the Future** (1999*), Sustainable Development education: Design Specification*, HMSO, London.

11. **Ali Khan** (1995), *Taking Responsibility: Promoting Sustainable Practice Through Higher Education Curricula*, Pluto Press, London.

12. **Design Council** (1997), *More for Less: Design for Environmental Sustainability*, Design Council, London.

13. **Fletcher K. and E, Dewberry** (2002), *demi: a case study in design for sustainability*, International Journal of Sustainability in Higher Education: 38-47, Emerald.

14. **Lilley, D.** (2003) *Xxxxxxxxx and Xxxxxxxx: Exploring perceptions of sustainability to inform communication strategies*, MRes Thesis, Cranfield University, Cranfield.

15. **Weizsäcker E., Lovins A.B., and L. H. Lovins,** (1997), *Factor Four – Doubling Wealth, Halving Resource Use*, Earthscan, London.

16. **Bohm, D.,** (Nichol L. Ed), (1998), *On Creativity*, Routledge, London.

Designing and Teaching a Sustainable Design Course at Purdue University

Julie Ann Stuart [1], Ph.D., P.E., Inez Hua [2], Ph.D.
[1]School of Industrial Engineering Purdue University West Lafayette, IN   47907-2023, USA
[2]School of Civil Engineering Purdue University West Lafayette, IN   47907-1284, USA

ABSTRACT

The Schools of Industrial Engineering and Civil Engineering teamed to develop a multi-disciplinary graduate course in Sustainable Design at Purdue University which was introduced fall 2003. In this paper, we show how to integrate industrial engineering and environmental engineering perspectives into a valuable sustainable design curriculum. References to sustainable design teaching materials and efforts are briefly summarized. The course curricula included design principles related to material selection and pollution prevention in process design, energy efficiency, consumer impacts, reuse, remanufacturing, and recycling. In addition, educational modules on product life-cycle assessment, environmental regulations, environmental costing, and corporate sustainability goals are discussed. Next, innovative teaching exercises to help students learn how to translate sustainable design concepts into practice are described. Student feedback on perceived changes in knowledge of sustainable design topics is reviewed. The paper concludes with recommendations for sustainable design curricula and teaching modules.

1 INTRODUCTION

Sustainable design requires the study of the interactions between industrial metabolism and environmental metabolism. Linkages between these two disciplinary perspectives are recognized when considering the industrial engineering study of industrial metabolism, the linkages between suppliers, manufacturers, consumers, refurbishers, and recyclers, and the environmental engineering study of environmental metabolism, the linkages between entities such as biota, land, freshwater, sea water, and atmosphere. Since industrial processes are part of a complex global web of activities related to raw materials extraction, manufacturing, assembly, distribution, sales, use, reuse, recycling, and disposal, it is important for engineers to learn how to assess the life cycle environmental impacts of their product and process designs. Sustainable design seeks to employ green engineering principles to use renewable resources (1). Sustainable design curricula is an important area of engineering and science education research (2, 3, 4, 5, 6, 7, 8, 9). The publication of textbooks in industrial ecology (10, 11), green engineering (12), design for environment (11, 13, 14, 15, 16, 17, 18, 19, 20, 21), environmentally conscious manufacturing (22, 23, 24, 25, 26, 27), and pollution prevention (28) support the development of courses that include sustainable design principles. These textbook topics are necessarily quite broad and include concepts from multiple engineering disciplines. Thus, a sustainable design course may include instruction from not only an environmental perspective, but also economic and manufacturing perspectives. To this end, the Schools of Industrial Engineering and Civil Engineering teamed to develop a multi-disciplinary graduate course in Sustainable Design at Purdue University which was introduced fall 2003. The course included the topics summarized in Table 1 by conceptual influence. Environmental engineering contributions to the course focused on assessing environmental impacts, earth systems, and environmental regulations while industrial engineering contributed design, materials, and economic objectives. By combining multiple disciplinary perspectives, the course more realistically exposes students to the complexity of sustainable design and the need for cooperation among disciplines. An important inclusion is technological and societal frameworks for environmental challenges.

Table 1. Purdue Sustainable Design Course Topics by Primary Conceptual Influence.

Environmental Engineering Conceptual Influence	Industrial Engineering Conceptual Influence
• Earth systems engineering management • Environmental process design and operations • Environmental interactions • Life cycle assessment • Sustainable goals and metrics • Environmental aspects of government/law	• Design for energy efficiency • Design for product delivery and service • Design for extended product life • Design for end-of-life • Engineering economics and environmental accounting • End-of-life product networks and process design

Table 2 provides a sample of courses with sustainable design content. Additional list of universities with pollution prevention and industrial ecology are available (29, 30). Several of the courses listed in Table 2 were listed with multiple course codes denoting their cross-disciplinary nature. The course at Purdue was coded for both civil and industrial engineering and open to all science and engineering graduate students. Next, we discuss how the course structure enhanced student learning.

D004/005/2004

Table 2. Sample of Courses that include Sustainable Design.

University	Course Name & Code	Reference
University of Michigan	Industrial Ecology CEE 586 (Nat. Res. 557)	University of Michigan, Civil and Environmental Engineering Course Bulletin 2003-2004; http://www.engin.umich.edu/students/current/academics/courses/cee.pdf. (accessed May 2004.)
Stanford University	Green Architecture CEE 136	Stanford University, Stanford Bulletin 2003-04, pages 126-140; http://www.stanford.edu/dept/registrar/bulletin/pdf/CivilEng.pdf. (accessed May 2004.)
University of California – Berkley	Industrial Ecology ER 290-4	University of California – Berkley, Environmental Sustainability courses; http://faculty.haas.berkeley.edu/toffel/workinggroup/ucb_courses.htm, (accessed May 2004, updated August 2002.)
Yale University	Greening the Industrial Facility F&ES 500a	Yale University, School of Forestry & Environmental Studies; http://www.yale.edu/environment/popup/courses/industrial_manage.html, (accessed May 2004, copyright 2001.)
Massachusetts Institute of Technology	Planning for Sustainable Development 11.366	Massachusetts Institute of Technology, Environmental Policy Group, Sustainable Development Specialization, http://web.mit.edu/dusp/EPG/academics/spec4.html, (accessed May 2004, copyright 2004.)
Chalmers University of Technology	Life Cycle Assessment	Chalmers University of Technology, Physics and Engineering Physics Industrial Ecology Program; http://www.fy.chalmers.se/edu/imp/ie.xml, (accessed May 2004.)
Cranfield University	Ecodesign	Cranfield University, Sustainability & Design option for Master's Manufacturing Program; http://www.cranfield.ac.uk/sims/ecotech/msc/s&dmsc.htm, (accessed May 2004, updated 2004.)
Universite de Technololgie Troyes	Industrial Ecology and Sustainability	Universite de Technololgie Troyes, Engineering Specialization; http://www.utt.fr/uk/catalogue/majorSpecializations.php?rub=03, (accessed May 2004.)
Norwegian University of Science and Technology	Systems for Recycling and Closed Material Loops IE5	Norwegian University of Science and Technology, Industrial Ecology Program; http://www.indecol.ntnu.no/indecol.php, (accessed May 2004, updated 2003.)
Carnegie Mellon	Environmental Life Cycle Assessment and Green Design CEE 12-714	Carnegie Mellon, Civil and Environmental Engineering Course Description; http://gdi.ce.cmu.edu/, (accessed May 2004.)
University of South Carolina	Sustainable Design and Development EMCH 529	University of South Carolina, Mechanical Engineering Graduate Student Handbook; http://www.me.sc.edu/grad/PDF/HANDBK03-04.pdf

2 COURSE STRUCTURE

Using educational modules, the course introduces students to product life-cycle assessment, environmental regulations, environmental costing, and corporate sustainability goals were developed. Modules were composed of a reading assignment, lecture notes, a group discussion question, and a homework exercise. The reading assignments were derived from the textbook, journal papers, conference papers, and websites. Lectures included both notes and a group discussion question to promote active learning. An example of an active learning exercise is given in Figure 1. Students formed small groups and recorded their responses; next, whole class interaction encouraged discussion and debate. These classroom exercises helped students learn how to translate sustainable design concepts into practice. One challenge was providing adequate time to introduce and carry out active learning exercises. As a result, introductory information for active learning exercises will be included in the reading assignments prior to classroom discussion in the future.

Lecture 18: Active Learning Worksheet

Topic: Environmental Metrics and Indicators

In groups of 3-4 students, discuss Table 12.2 in Graedel, T. and B. Allenby, Chapter 12, "How Green is the Automobile and Its Infrastructure?" *Industrial Ecology of the Automobile*, Prentice Hall, 1998. p. 168. (31)

- Which option has the highest environmental impact during the
  production stage?
  product use stage?
  waste stage?
  overall?
- What assumptions did you make in answering the previous question?
- What is the impact of the materials extraction stage?
- What other alternatives might an engineer investigate?
- Comment on the EPS metrics.

Figure 1. Example of an Active Learning Exercise for the Purdue Sustainable Design Class

Homework exercises required assessment and application of the topics listed in Table 1. An example of a portion of one homework exercise is given in Figure 2. The homework assignments were intended to provide students the opportunity to critically analyze a complicated design issue, and then apply various course topics to arrive at a solution. Furthermore, the homework assignments were meant to illustrate the relevance of sustainable design principles to a wide variety of engineering disciplines. Homework assignments were individual effort (e.g., each student turned in the assignment and received a grade); however, students were encouraged to discuss the assignments among themselves.

Assignment 4

Read: Blanchard, S., and Reppe, P., "Life Cycle Analysis of a Residential Home in Michigan" (University of Michigan, Report No. 1998-5, September 1998)

Q1:
a) Describe the functional unit(s) for comparison.

Q2:
a) Which life-stages of a residential home were included in the study?
b) What was the approximate "temporal boundary" for the system that was being analyzed?
c) Construct a diagram indicating inputs and outputs at the various stages included in the study.

Q3:
The life cycle global warming potential was estimated for each home. For which processes or activities were GWP emissions estimated

Figure 2. A portion of a homework assignment for the life cycle assessment module of the Purdue Sustainable Design course

The success of the course structure was assessed through student feedback.

3 STUDENT FEEDBACK

Student feedback on perceived changes in knowledge of sustainable design topics was collected through a survey at the end of the course. The survey asked students to evaluate their knowledge of the course material based on a three level scale represented by "new topic," "limited knowledge," and "knowledgeable." The course survey is provided in Table 3. The survey response rate was 70%; 17 out of 24 students completing the course responded at the end of the fall 2003 semester. The Purdue course was an elective (non-required) graduate course.

Table 3. Sustainable Design Course Survey and Abbreviations.

Please place a B (before) and an A (After) in the columns that correspond to level of knowledge before and after taking this course.			Concept	
1 Know-ledgeable	2 Limited knowledge	3 New topic	Description	Abbreviation
			Sustainable design	Sustainable design
			Material selection with respect to material consumption and world supplies	Material selection
			Pollution prevention in manufacturing process design	Pollution prevention
			Design for energy efficiency	Design for energy efficiency
			Design to reduce environmental impacts from consumer use	Design for consumer use
			Design for reuse	Design for reuse
			Design to reduce environmental impacts at end-of-product-life	Design for recycling
			Product life cycle assessment	Product LCA
			Environmental regulations	Environmental regulations
			Environmental costs	Environmental costs
			Measuring corporate sustainability performance	Corporate sustainability performance
			Corporate sustainability goals	Corporate sustainability goals

Using the abbreviations in Table 3, Figure 3 summarizes the students' perceptions of their knowledge of the course content before the course and at the end of the course. The results for knowledge prior to the course reflect that some of the students in the course were engaged in research related to environmentally conscious product and process design, environmental remediation, environmental management, and reverse production systems while other students were introduced to course topics for the first time. The course significantly increased knowledge in all areas, especially in the areas of corporate sustainability, product life cycle assessment, design for recycling, design for reuse, and design for consumer use. Areas where course c ontent c ould b e i ncreased include environmental regulations and design for energy efficiency.

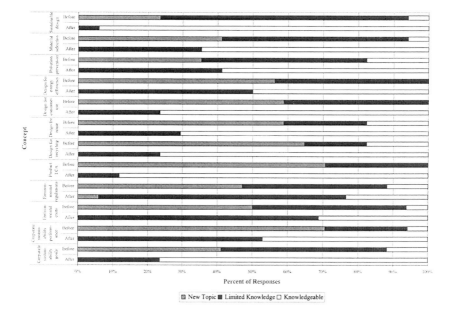

Figure 3. Student Feedback for Purdue Sustainable Design Course

4 CONCLUSIONS

Sustainable design is primarily isolated to specific textbooks and specific courses. The integration of multiple disciplines is important to sustainable design curricula. Course structure that includes active learning exercises, class discussions, and homework assignments are also important to the level of student learning. In the future, the integration of these concepts throughout engineering curricula will enable sustainable design concepts to permeate the approaches and technology developed and utilized by future engineers.

5 ACKNOWLEDGEMENTS

The authors thank Lisa Tieman for graphing the student responses in Figure 3. This work was supported in part by the National Science Foundation (Award Number 023061). Any opinions, findings, and conclusions or recommendations expressed in this material are those of the authors and do not necessarily reflect the views of the National Science Foundation.

REFERENCES

1 **Anastas, P. T. and J. B. Zimmerman** (2003) Design through the 12 principles of green engineering *Environmental Science & Technology* **37**(5): 94A-101A

2 **Friedlander, S. K.** (1989) Environmental Issues: Implications for Engineering Design and Education In J. H. Ausubel and H. E. Slodovich (ed.), *Technology and Environment* (Washington, DC, National Academy Press, pp. 167-181.

3 **Ehrenfeld, J. R.** (1994) Industrial Ecology and Design for Environment: The Role of Universities In B. R. Allenby and D. J. Richards (ed.), *The Greening of Industrial Ecosystems* (Washington, DC, National Academy Press, pp. 217-227.

4 **Cooper, J. S. and J. Fava**, (1999) Teaching Life-Cycle Assessment at Universities in North America *Journal of Industrial Ecology* **3**(2&3): 13-17.

5 **Marstrander, R., H. Brattebo, K. Roine and S. Storen**, (1999) Teaching Industrial Ecology to Graduate Students *Journal of Industrial Ecology* **3**(4): 117-130

6 **Brattebo, H.** (2001) Industrial Ecology and Education *Journal of Industrial Ecology* **5**(3): 1-2.

7 **Cooper, J. S. and J. Fava** (2000) Teaching Life-Cycle Assessment at Universities in North America, Part II *Journal of Industrial Ecology* **4**(4): 7-11.

8 **Mihelcic, J. R., J. C. Crittenden, M. J. Small, D. R. Shonnard, D. R. Hokanson, Q. Zhang, H. Chen, S. A. Sorby, V. U. James, J. W. Sutherland and J. L. Schnoor**, (2003) Sustainability Science and Engineering: The Emergence of a New Metadiscipline *Environmental Science & Technology* **37**: 5314-5324.

9 **Shi, H., Y. Moriguichi and J. Yang** (2003) Industrial Ecology in China, Part II: Education *Journal of Industrial Ecology* **7**(1): 5-8.

10 **Bourg, D. and S. Erkman**, Eds. 2003. *Perspectives on Industrial Ecology.* (Sheffield: Greenleaf Publishing).

11 **Graedel, T. E. and B. R. Allenby** (1996) *Design for Environment.* (Upper Saddle River, NJ: Prentice Hall, Inc.)

12 **Allen, D. T. and S. D. R.** (2002) *Green Engineering: Environmentally - Conscious Design of Chemical Processes.* (Upper Saddle River, NJ: Prentice Hall).

13 **Congress, U., Office of Technology Assessment** (1992) *Green Products by Design: Choices for a Cleaner Environment*, report OTA-E-541.

14 **Fiksel, J.** (1993) *Quality Metrics in Design for Environment*: Decision Focus Incorporated).

D004/005/2004

15 **Keoleian, G. A., D. Menerey and M. A. Curran** (1993) *Life Cycle Design Guidance Manual*. (Cincinnati, OH: EPA).

16 **Keoleian, G. A., D. Menerey, B. W. Vigon, D. A. Tolle, B. W. Cornaby, H. C. Latham, C. L. Harrison, T. L. Boguski, R. G. Hunt and J. D. Sellers** (1994) *Product Life Cycle: Assessment to Reduce Health Risks and Environmental Impacts*. (Park Ridge, NJ: Noyes Publications).

17 **Behrendt, S., C. Jasch, M. C. Peneda and H. v. Weenen** (1997) *Life Cycle Design: A Manual for Small and Medium-Sized Enterprises*. (New York: Springer-Verlag).

18 5 **Billatos, S. B.** (1997) *Green Technology and Design for the Environment*. (Storrs: Taylor and Francis).

19 **Keoleian, G., K. Kar, M. M. Manion and J. W. Bulkley** (1997) *Industrial Ecology of the Automobile: A Life Cycle Perspective*. (Warrendale, PA: Society of Automotive Engineers, Inc.).

20 **Smet, B. D.** (1998) *Using LCA in Environmental Decision-Making*. (Dordrecht: Kluwer Academic Publishers).

21 **Lewis, H., J. Gertsakis, T. Grant, N. Morelli and A. Sweatman** (2001) *Design + Environment*. (Sheffield: Greenleaf Publishing).

22 **Cattanach, R. E., J. M. Holdreith, D. P. Reinke and L. K. Sibik** (1995) *The Handbook of Environmentally Conscious Manufacturing*. (Chicago: Irwin Professional Publishing).

23 **Roberts, P.** (1995) *Environmentally Sustainable Business*. (London: Paul Chapman Publishing, Ltd.).

24 **Thomas, S. T.** (1995) *The Facility's Manager's Guide to Pollution Prevention and Waste Minimization*. (Washington, DC: The Bureau of National Affairs, Inc.).

25 **Epstein, M. J.** (1996) *Measuring Corporate Environmental Performance: Best Practices for Costing and Managing an Effective Environmental Strategy*. (Chicago, IL: Irwin Professional Publishing).

26 **Theodore, M. K. and L. Theodore** (1996) *Major Environmental Issues Facing the 21st Century*. (Upper Saddle River, NJ: Prentice Hall PTR).

27 **Sarkis, J.**, Ed. 2001. *Green Manufacturing and Operations: from Design to Delivery and Back*. (Sheffield, UK: Greenleaf Publishing Ltd.).

28 **Allen, D. T. and K. S. Rosselot** (1997) *Pollution Prevention for Chemical Processes*. (New York: John Wiley & Sons, Inc.).

29 **Cockrill, K**. (2004) Industrial Ecology in Higher Education [online] available at http://www.is4ie.org/academic_programs.html (20 May 2004)

30 National Pollution Prevention Center for Higher Education (1995) [online] available at http://es.epa.gov/techinfo/facts/nppche.html (20 May 2004)

31 **Graedel, T. E. and B. R. Allenby** (2003) *Industrial Ecology*: Prentice-Hall).

The EcoDesign co-operative – a global initiative to enhance EcoDesign education through courseware exchange

W DEWULF, D CLEYMANS, and **J DUFLOU**
Department of Mechanical Engineering, Katholieke Universiteit Leuven, Belgium
J JESWIET
Department of Mechanical Engineering, Queen's University, Canada

ABSTRACT

EcoDesign Education is a key factor in shaping the essential consciousness, knowledge and skills to design a sustainable future. The decision-making skills to be taught in sustainable design courses imply the ability to consider expanded time and space horizons and to examine cross-disciplinary effects. To enhance the inter-cultural and inter-disciplinary character of EcoDesign education, a global EcoDesign Cooperative has been founded, providing a platform for sharing EcoDesign course material. Material exchange occurs electronically through the EcoDesign Cooperative web-portal, introduced in this paper.

1 INTRODUCTION

Over the past decades, awareness has grown that end-of-pipe emission treatment alone will not be able to sufficiently reduce the environmental decline. A more holistic approach is needed, which enables to reduce the environmental burden per unit produced by an order of magnitude. Such far-stretching reduction can only be achieved while considering the product's life cycle environmental impact from the early design stages onwards.

Intensive research programs have led to the development of numerous EcoDesign tools and procedures. Meanwhile, pilot projects have demonstrated the applicability of EcoDesign principles. Another key factor for making EcoDesign happen in daily industrial practice is, however, the education of current as well as future designers, engineers and managers. Providing an appropriate understanding of environmental threats and of technological opportunities shapes the essential consciousness, knowledge and skills to design a sustainable future. *If you are thinking one year ahead, sow seeds. If you are thinking ten years ahead, plant a tree. If you are thinking a hundred years ahead, educate people.* By dedicating 2005-2014 as the Decade of Education for Sustainable Development [1], the United Nations have emphasized their sharing this vision.

2 THE FORMATION OF AN ECODESIGN COURSEWARE COOPERATIVE

Over the last few years, universities and high schools are increasingly integrating sustainability in the curricula of, for example, designers and engineers. Nevertheless, systematic teaching of EcoDesign at an undergraduate level is still lacking. Moreover, the decision-making skills to be taught in sustainable design courses imply the ability to consider expanded time and space horizons and to examine cross-disciplinary effects [2]. Current professional education is, however, often strongly oriented towards specialisation and knowledge rather than towards cross-disciplinarity and skills. EcoDesign education therefore implies a challenge, willingly taken up by numerous teachers worldwide. Their interest in special journal issues on sustainability education ([3], [4]) and in dedicated conferences ([5]) indicates their recognition of the need to exchange views, concepts and material to realise this global cross-disciplinarity.

From the same perspective, the EcoDesign Cooperative was founded by leading EcoDesign teachers from Europe, Canada, the US, and Australia in order to facilitate inter-disciplinary, inter-institutional, and inter-cultural exchange of EcoDesign course material. For this purpose, course material can be defined rather loosely as material which can be used for educational purposes, including full course descriptions, lecture slides, reports, graphs, etc.

The cooperative was founded in the framework of the CIRP working group on Life Cycle Engineering, which organises the yearly International Seminar on Life Cycle Engineering [6]. Current members of the EcoDesign Cooperative include Jack Jeswiet (Canada), Joost Duflou (Belgium), Conrad Luttropp (Sweden), Günther Seliger (Germany), Bert Bras (US), Leo Alting (Denmark), Walter Olson (US), Hartmut Kaebernick (Australia), Rolf Steinhilper (Germany), Delsey Durham (US). However, membership is open to anyone willing to actively share course material with the other members.

3 A WEB-PORTAL FOR ECODESIGN COURSEWARE EXCHANGE

Material exchange occurs electronically through the EcoDesign Cooperative web-portal at http://www.mech.kuleuven.ac.be/eco/ (Figure 1). While part of the material is accessible to the wider public, another part remains exclusively accessible to EcoDesign Cooperative members, who are engaged to provide input to the courseware base.

As stated above, EcoDesign course material can be defined rather loosely as material that can be used for EcoDesign educational purposes. Next to lecture handouts and slides, a large variety of information sources such as internet links, research reports and articles are used in academic education in order to keep course information fresh and up to date. Due to the difference of pedagogic approaches followed by individual teachers, background information for one course might be used as principal lecture material in another course, or vice versa. Therefore, the initiators have declined from meticulously predefining limitations or classifications for the material to be uploaded. Moreover, the sources of educational information are available in a large variety of media such as .pdf, .ppt, .doc, .gif, and .jpg formats. Therefore, the web-portal has preserved full flexibility of uploading any type of educational material without the need for very rigid document classification.

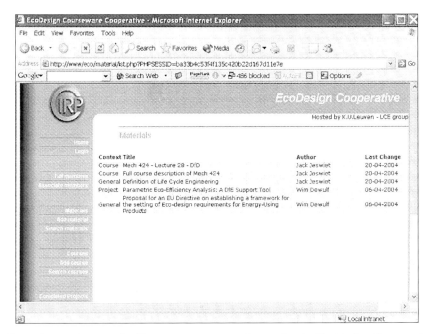

Figure 1. A web-portal facilitates courseware exchange for the EcoDesign Cooperative

Courseware sharing implies more than exchanging lecture contents and background information. Therefore, a second part of the web-portal aims at sharing experience on EcoDesign course approaches: how is my course organised, what are the objectives, what is the approach, in which educational context? Courses can be linked to any number of course materials entered under the first part of the web portal.

Data from research is often used in academic education. Moreover, EcoDesign teachers are most often also active in the related research field. Third, many EcoDesign courses are taught in an interactive way, in which students can prove their knowledge and exercise their skills in small-scale EcoDesign projects. Therefore, a third part of the web-portal allows a listing of research projects by EcoDesign Cooperative members. By providing the option to propose also new project ideas, the Cooperative members can benefit from this cross-disciplinary environment, not only for their educational work, but also for their research activities.

A list of members and access to their websites concludes the contents of the web-portal. By allowing links between the different information items, it is at any time possible to retrieve the lecture material used for a particular course, or courses and projects supervised by a particular member (Figure 2).

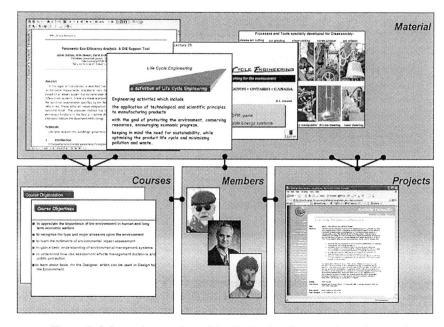

Figure 2. Schematic overview of the EcoDesign Cooperative web-portal

4 ADVANTAGES OF THE ECODESIGN COOPERATIVE

The web-based exchange of course material implies a number of advantages for the EcoDesign Cooperative members as well as for the development of (Eco)Design curricula.

First, EcoDesign is still a young discipline, which is evolving very rapidly. Due to its interdisciplinary character, EcoDesign teachers need to keep track of advances in many domains, ranging from environmental problems, energy conversion, recycling technology, advanced control techniques, cleaner production and green chemistry to design science, supply chain management and reverse logistics. When analysing the professional background of EcoDesign teachers, the same diversity in research interests can be distinguished. Consequently, intensive interdisciplinary cooperation allows keeping teaching material up to date by exchanging course material among field experts of different domains.

Second, the sensitiveness and attitude with respect to the different environmental problems differ amongst regions and continents. This is translated into differing legislative actions around the globe. Therefore, a global cooperative allows being aware of the important evolutions in different jurisdictions.

Third, many EcoDesign courses are web-based. Members of the EcoDesign Cooperative have expressed their willingness to contribute to a common, international course on EcoDesign. Consequently, each teacher can take care of a module concentrating on his own core competences, while making use of other modules provided by highly esteemed colleagues.

Finally, as stated above, the cooperative can lie at the basis of new project ideas to enhance not only EcoDesign education, but also EcoDesign research.

5 CONCLUSIONS

EcoDesign is, by its nature, cooperative and not competitive. Therefore, the newly founded EcoDesign Cooperative provides a framework to facilitate inter-disciplinary, inter-institutional, and inter-cultural exchange of EcoDesign course material. Material exchange occurs electronically through the EcoDesign Cooperative web-portal at http://www.mech.kuleuven.ac.be/eco/.

REFERENCES

1 **UN.** (2002) Resolution 57/254 on the United Nations Decade of Education for Sustainable Development (2005-2014).

2 **Hodge, T.,** and **Nesbit, S.** (1999) Moving towards sustainability: the new professional imperative. In *Innovation Journal of the Association of Professional Engineers and Geoscientists of BC*, 3(6), pp. 16-7.

3 **N.** (2000) Journal of Cleaner Production 8, Special Edition on "Education, Training, and Learning".

4. **N.** Journal of Cleaner Production, Special Issue on "Education for Sustainable Development", forthcoming.

5. **N.** (2002) Engineering Education for Sustainable Development, Delft, October 24th-25th, 2002.

6. **Alting, L.** (2003) Engineering for Sustainable Development – An Obligatory Skill of the Future Engineer, Proceedings of 10th CIRP International Seminar on Life Cycle Engineering, Copenhagen, 21-23 May 2003.

Methods and Tools for
Sustainable Design

Beyond the Eco-Functional Matrix – design for sustainability and the Durham methodology

T D SHORT and **C A LYNCH**
School of Engineering, University of Durham, UK

ABSTRACT

At the 2003 International Conference on Design and Manufacture for Sustainable Development, Jeswiet [1] presented a keynote paper which included considerable information from Jessica Lagerstedt's PhD thesis [2]. Lagerstedt considered a wide variety of eco-design tools, their strengths and weaknesses and produced a new design tool – the "Eco-functional matrix" – to aid designers early on in the design process, when the most changes can be made. This paper reports on work carried out at the University of Durham to extend Lagerstedt's investigation. It describes the use of Lagerstedt's Eco-functional matrix (EFM) on a number of products at different stages in their life cycles. It goes on to analyse the results of these trials, identifying the strengths and weaknesses of the EFM as a design tool. The paper then illustrates how the EFM has been adapted and improved, bearing in mind the lessons learnt during the initial trial, in order to fit in with the Durham "team based" design methodology. Finally the paper considers the results of the application of the new tool on a further product.
The paper therefore proposes a new design tool, usable throughout the product design process, and shows how it provides a clearer focus for designing sustainability into a product.

1 INTRODUCTION

Environmental sustainability is becoming increasingly important for a company both economically and socially. Bjorn Stigson of the World Business Council for Sustainable Development, states *'... it is not clear to what extent consumers are willing to pay a premium for green products. But they can certainly be quick to turn away if companies are seen to be ignoring such concerns.'* [3]. Many environmentalists argue that sustainability is becoming a selling point, but are consumers willing to pay? Irrespective, if a product is 100% sustainable but does not fulfil its intended purpose and function then it will not be used. It will be a waste of resources. Product functionality must not be compromised to increase sustainability.

This report looks at how sustainability and functionality can be incorporated into product design by examining Lagerstedt's design tool – The Eco-Functional Matrix (EFM) – testing it on various products and examining the results. The aim is to find a tool that is simple to use and provides sufficient information for a designer to create products that are both functional and sustainable.

2 DESIGN AND SUSTAINABILITY

For many companies, the definition of sustainability is to survive [4]. This is not a future aim for a company but it is something that must be considered now. If a company is not able to sell its products, at a profit, then it will not survive. However if all a company focuses on is the here and now, then it cannot have a certain future.

There are many design methodologies that allow the consideration of factors affecting a product, such as Quality Function Deployment, Design for Assembly and Design for Manufacture. Short et al [4] showed that, although not always recognised as sustainability tools, such methodologies can play a significant part in the sustainability of a company's business ("Industrial Sustainability") as well as the more widely understood "Brundtland Sustainability". Other tools, such as Life Cycle Analysis, are known to directly address Brundtland Sustainability. Some of these methodologies are further investigated by Lagerstedt [2] and found to be wanting, which leads her to develop her Eco-Functional Matrix.

2.1 The Eco Functional Matrix

Lagerstedt addresses the requirement that sustainability be considered early in the design process. By comparing existing design tools and studying the issues involved in sustainability she proposes the EFM as a way to enable designers to consider the sustainability of their product from its conceptual design stage throughout the entire process.

The EFM (see **Figure 1**) is divided into two areas of analysis, the functional profile and the environmental profile of a product. These two areas then combine to form the matrix.

A variable speed drive integral inverter and motor for driving a variety of loads. User: Typically manufacturers of mechanical handling equipment	Value	A) Lifetime	B) Use-time	C) Reliability	D) Safety	E) Human/machine	F) Economy	G) Technical Flexibility	H) Environmental demand	K) No. of products/year	L) Size	M) No. of different materials	N) Material mix	O) Scarce material	P) Toxic material	Q) Energy	R) Energy sources
A) Lifetime	6																
B) Use-time	7																
C) Reliability	8																
D) Safety	7																
E) Human/machine	4																
F) Economy	7																
G) Technical Flexibility	4																
H) Environmental demand	6																
K) No. of products/year	3	X	X				X		X								
L) Size	5						X		X								
M) No. of different materials	6		X				X										
N) Material mix	8	X		X			X										
O) Scarce material	5		X				X		X								
P) Toxic material	3	X				X			X								
Q) Energy	6	X	X						X								
R) Energy sources	8																

Figure 1 - Example of EFM

2.1.1 Functional Profile

The Functional Profile ("FP": categories A – H) is primarily based on criteria relating to user benefit. It establishes what the customers require of the product, and how important each of these requirements is. The aim of the FP is to establish 'what are the functional priorities of the product and what is the lowest environmental cost we can incur whilst obtaining them?'. See [2] for a full description of these categories.

2.1.2 Environmental Profile
The Environmental Profile ("EP": Categories K – R) evaluates the impact of the product on the environment. The areas to be considered have been chosen to enable identification in early design phases and are quantitatively assessed according to Lagerstedt's criteria and therefore given a value between 0 and 10, 10 representing the most dominating environmental impact (for a full description, see [2]).

2.1.3 Creating the Matrix
The values generated in the FP and E P are transferred onto the matrix in the first column labelled Value. The next step is to establish whether there is any interrelationship between the categories in both the EP and the FP. For example, does the safety of a product have any relation to the size of the product? Any relationships are marked on the matrix with a cross in the correlating square. **Figure 1** shows an example of a completed EFM.

Area 1 of the matrix helps to identify any important functional correlations, and Area 3, the environmental correlations.

Area 2 of the matrix indicates the area of interest as it demonstrates how the environmental and the functional aspects are linked.

Categories that have been assigned a value of 5 or more are considered to be important factors in the development of the product and are highlighted in bold. If two categories with high values show an interrelationship, this is highlighted by a bold cross at the corresponding link. The bold cross indicates a 'critical point'. Lagerstedt states that these areas should be discussed at length.

3 USING THE EFM

The EFM was tested on three products:
    An Integral Motor-Inverter (FKI Industrial Drives).
    A "Pipeline Tractor" for internal pipe inspection (Durham Pipeline Technology).
    A "Beer Widget" to produce a good head on beer from a can (Cambridge Consultants).
Each test took between 35 and 40 minutes to carry out.

3.1 Analysis of EFM
The results of these tests are not themselves presented in detail here, as it is more important for this paper to consider the strengths and weaknesses of the methodology.

The range of products on which the EFM was carried out show a range of different levels of EP scores. Apart from the 'Number of products produced/year', the beer widget had very low values for all EP categories whereas the motor produced much higher values, the majority being 5 or higher. The EP for the pipeline tractor is between the two.

3.1.1 Layout
The EFM is simple to fill in, with a simple structure and clear procedure, keeping completion time for the EP short. The simplicity of the matrix means that it does not deter a designer from trying to fill it in and the information provided is easy to interpret. If analysing the redesign of an existing product then the FP is typically identical to the previous one. For New Product design, gathering information on customer needs and requirements may take longer,

but should be carried out whether or not the EFM is to be used. For all of the tests carried out this information was available, and the EFM was completed within 45 minutes.

3.1.2 Scoring Methods
Whilst carrying out the tests it became clear that the numerical values for the environmental questions were particularly useful. Environmental issues are often seen as qualitative, which could make it more difficult to improve substantially a product's design. With the EFM, however, there are definite answers to the questions in the Environmental Profile, which results in a set score for the categories.

3.1.3 Product Comparison
The EFM provides no comparison to competitors' products. The motor-inverter tested is of higher efficiency than standard, so could be considered a more sustainable design. There was no area of the matrix where this was taken into account. However, it is not always good to compare one product with another as this can inhibit creativity and new thinking.

'By moving away from the 'how can we be less bad?' mentality to the 'how can we be 100% good' mindset, we give ourselves the capability of redesigning every product to be 100% sustainable.' [5]

3.1.4 Relationships
The crosses in Section 2 of the diagram show the interrelationship between the categories in the functional and environmental profiles. They do not, however, indicate whether the relationship is a positive or a negative one. For example, the results for the beer widget show that there is a relationship between categories D and P (safety and toxic material). In this case, in order to make the product safe, no toxic material is used, however the cross provides no explanation of this. Similarly, toxic material could be used to ensure safety of the product. The difference between these two situations would alter the interpretation of the results. The crosses are quick to fill in but often they do not provide all of the information that is required.

3.1.5 Prioritisation of Changes
In the EFM Lagerstedt proposes highlighting the crosses that show relationships where both EP and FP scores are high. These are the areas that should be addressed first with regard to making changes in the design. In practice, where FP scores are high, the particular function of the product is important and can rarely be compromised, therefore making changes will be much harder. It is easier to make changes in areas where only the environmental score is high. Looking at these first will allow prioritisation of the areas where the most drastic changes can be made, with the least effort. It is equally just as important to look at areas where the environmental score is high regardless of the functional score, as the latter may not alter but the environmental score may. It is the reduction of environmental scores that will result in a product becoming more sustainable.

3.1.6 Manufacture
The EFM considers energy used during the life time of a product but not the energy involved in making it – in fact no account at all is taken of the manufacturing process. If a product is 100% sustainable but requires valuable resources for its manufacture, then it is not a good product. Recognition of the need for cleaner production by the United Nations Environment Programme [6] emphasises the requirement for it to be included in a "design for sustainability" tool.

3.2 Summary

The results of the tests using the EFM have shown a number of strengths and weaknesses. Despite some significant shortcomings, the EFM provides a good starting point for the analysis of a product's sustainability.

4 THE "DFSM"

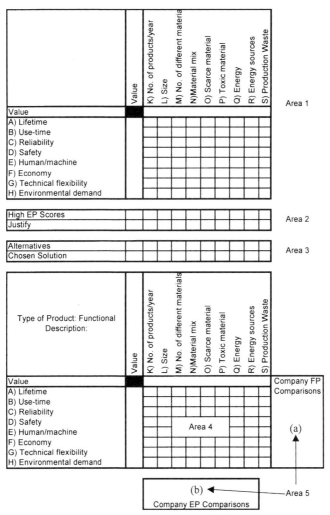

Figure 2 - Layout of the DFSM

A new matrix – the Design for Sustainability Matrix (DFSM) – has been designed based on the EFM. Where the original matrix finishes, the DFSM continues throughout the redesign process. The aim is to encourage a project team to develop the points raised by the methodology, the momentum generated by the initial matrix therefore being carried on to make the new design more sustainable.

4.1 Team Based Design for Sustainability

The methodology has deliberately been developed to promote teamwork, the benefits of which have been documented [7, 8] and are assumed for this paper. The matrix is now formed on an A0 acetate-type backing sheet, using "Sticky-notes" such as Post-it® notes. This removes the reliance on a computer, allows uninhibited group discussions, promotes creative expression and allow ideas to be moved around or disposed of if unsuitable.

4.2 The Design for Sustainability Matrix (DFSM)

4.2.1 Breakdown of Layout

The DFSM in **Figure 2** consists of 5 areas. Area 1 of the matrix is based upon the original EFM, but using only the 2nd area of the EFM, where the EP and the FP are compared. The matrix has also been rotated 90° so the EP is along the top instead of down the sides. The addition of an EP category 'S' involves the manufacturing process of the product and is considered below.

Instead of placing crosses on the areas of interrelationships within the matrix, Sticky-notes are used, on which comments about the relationships are written. The notes explain why they are related, whether they are positive or negative relationships and how strong the relationship is.

Area 2 of the DFSM highlights elevated EP scores. As a guide, a score of 5 could be considered to be a high score, although a lower figure is clearly preferable. The letter of the category is repeated under its original position and the score repeated underneath it. This separation of these 'high scoring' categories from all of the EP categories allows priority to be given to the more important areas. It also allows the designers to focus on only these areas without being distracted by the other categories.

Once the Sticky-notes have been put in place they are then explained. Questions asked should include:

What makes the score so high?

Why is this used?

What are the benefits of using that method/material?

The third step is to look for possible alternatives. Some questions to ask would be;

Are there alternative materials that can be used?

Can a material be eliminated?

Can the process of manufacturing be altered?

These ideas should all be recorded on Sticky-notes, discussed and, where necessary researched. This section will be the most time consuming because it involves not only the generation of new ideas but ultimately the redesign of the product and the manufacturing process.

Whilst these ideas are being generated and developed the effect that they will have upon the Functional Profile must be considered – the new designs must not happen at the expense of the function of the product.

The DFSM is completed again in Area 4, but this time with the new scores that have been produced by the redesign. If the redesigns have been successful the EP values will be lower but the FP values should not have altered from the original matrix in Area 1. Again comments are written on Sticky-notes in the areas of interrelationships.

Area 5 allows comparisons with similar products from other companies. Section 5(a) compares their Functional Profiles and 5(b) their Environmental Profiles. For each of the categories, competitors' products are compared to the product that has just been designed. The products' names are written on Sticky-notes, one on each note and are then ranked with the best product on the inside of the matrix and the worst on the outside. This allows a direct comparison with other products, both for function and sustainability. Having the comparison at the end of the process ensures that the team re-design the product to be the 'best that it can be' rather than just 'better than the rest'.

4.2.2 Category S – Production Waste
A new category has been added to the Environmental Profile. Category S is titled Production Waste. The EFM had no way to measure how sustainable the manufacturing process was so category S aims to solve this problem. This category measures waste as the 'percentage waste based on total material cost', including the cost of energy, water and the raw materials. The pharmaceutical company Reckitt Benckiser, based in Hull have won a best practice award for their waste reduction programme, going from 4.1% to 2.6% waste based on total material cost, over a 12 month period to become one of the leading companies in minimising waste [9].

The scores for category S are as follows:

S   Production Waste

Percentage Waste based on total material cost.

0	0%
1	0 – 2%
2	2 – 4%
3	4 – 6%
4	6 – 10%
5	10 – 15%
6	15 – 20%
7	20 – 25%
8	25 – 30%
9	30 – 35%
10	>35%

5 USING THE DFSM

The DFSM was tested on a new domestic appliance designed by Cambridge Consultants. Areas 1 and 2 from **Figure 2** were completed over a period of 2 hours, at which point time constraints prevented the completion of the test.

5.1 Analysis of the DFSM
Again the specific results of using the DFSM will not be considered, this paper presenting an analysis of the application of the methodology.

5.1.1 Layout
Area 1 of the DFSM has a similar layout to the EFM and is again easily completed, allowing relationships to be shown clearly. Rewriting the Environmental Categories in Area 2 does allow a prioritisation in thought processes. It clearly separates the high scoring categories, letting designers concentrate only on these problems.

The DFSM is much larger than the EFM. Although the initial matrix is smaller because it only consists of the EP/FP comparison area, when notes are written on the interrelationships there needs to be more space.

5.1.2 Use of Sticky-Notes

The use of Sticky-notes encouraged discussion within the group, allowing every team member to write down their opinions even if this did lead to a repetition of points. The notes were big enough for the whole group to see what was written on the grid.

The different ideas that were put forward within the team started discussions which then brought forward other points. There were also a number of opinions to be considered which did not always agree so the discussions often lasted for a while.

5.1.3 Category S

Category S proved to be the hardest category to measure as it involved quite a bit of research. Power bills and the cost of inventory had to be found, which was time consuming but once discovered it was possible to measure the percentage of waste.

The only relationship between the Production Waste and the Functional Categories was the economy. Whether this is common to all products, rather than simply the product under test, is not clear but it seems quite possible. However, further research would need to be carried out by testing the DFSM on a wider range of products.

The category did provide a measure of the efficiency of the production process which allowed it to be compared with other products.

5.1.4 Notes on Interrelationships

Writing notes on the interrelationships provides significantly more information than simply placing crosses on the matrix. Thinking about how and why the areas are connected encourages discussion. Such discussion allows the designers to appreciate how altering one aspect of the product can affect other areas. Although this is more time consuming than just placing crosses in the corresponding areas, the extra information provided is very useful for later stages. The notes highlight the issues that should be addressed in the redesign process.

5.1.5 Incorporating Redesign into the DFSM

In the test that was carried out it was not possible to test the area that involved the redesign of the process. One of the benefits of the EFM was that it was quick to complete and so did not deter designers from finishing it. The addition of the redesign area in the DFSM takes away from the simplicity of the process.

However incorporating redesign encourages the thought processes to continue. Having the space to write down ideas from a group of people allows ideas to be generated whilst the environmental issues are still fresh in their minds. This encourages the continuation of the process.

5.1.6 Product Comparison

The product comparison area of the DFSM was not completed because the redesign did not take place.

6 GENERAL ANALYSIS

6.1 EFM

The main benefit of the EFM was that it was simple to complete and did not take a lot of time. The method was very straightforward, with most of the categories were easy to understand and the scoring bands were clearly defined. There appeared to be a tendency for the designers to exaggerate the sustainability of their design, but the quantitative environmental analysis did not allow that to happen.

The functional categories covered all of the major requirements of a product. They provided information about what the customers' requirements were and how important these factors are to the success of the product.

All of the factors that affect the sustainability of a product are covered in the Environmental Categories. The recyclability is addressed in the material mix category, the consumption of resources in the weight and number produced a year category, and there is the energy consumption category.

The EFM identifies possible problem areas, but does not always provide enough information to enable the designer to increase the sustainability of the product. Once the problem areas are identified the process is finished. There is also no account taken of the method of manufacture, which can have a large effect on the impact the product could have upon the environment.

Whilst carrying out the testing of the EFM some negative comments were made about the choice of Environmental Categories. It was noticed that the categories dealing with the materials used were all very similar and that there was a lot of overlap whilst discussing the relationships that occurred. The number of different materials used and the material mix were easily confused. This often led to confused discussions within the group. There was also some ambiguity in all of the tests about what was meant by the "Environmental Demand" category.

6.2 DFSM

The DFSM was created by adapting the EFM to create a more team based approach to the methodology. The method of using Sticky-notes was taken from the Durham Methodologies for DFM/DFA and other design approaches, to provide a way for a group of people to share their ideas. The method worked well as it allowed all the members to contribute. The new methodology also incorporates the method of company comparison that is used in Durham's Voice-of-the-Customer/Voice-of-the-Engineer methodology, a simpler alternative to QFD.

It was not possible to test the DFSM fully as it took a lot longer to complete than anticipated. Writing the notes rather than just placing crosses in the corresponding box provided much more information and indicated that the extra time taken is worth it. This added information should then be useful in the later stages of the process, when the redesign is being considered, enabling designers to make a more informed choice about what changes to make to increase the sustainability of a product. Having to explain why one area is related to another begins the process of connecting the functional and environmental categories. However one of the benefits of the EFM was that it was a quick and simple process.

The addition of Category S into the Environmental Profile permits consideration of the manufacturing process in the analysis. Measuring it as a percentage waste based on the total material cost was a simple measurement, if all the relevant information was available. It also means that not only the material waste is considered but also the consumption of energy and water. However this category does not really relate to any of the Functional Categories other than the cost of the product. Whether this is only true for the product tested is not clear but it seems likely that it is not. The way that a product is manufactured does not really relate to the customers demand of its functions. It also does not explain what the waste is, or where in the production process that it is occurring.

Discussions about possible alternatives that could be used were caused by the justification of why high scores in some environmental categories, but there was not enough time to research the suggested alternatives. There was, however, a momentum created to look into redesign. It seemed to be a natural progression from justifying why something was used, to arguing that an alternative material/ method might be as good but have a smaller environmental impact.

A comparison of competitors' products is also in the DFSM but has not been tested. It is expected that completing the methodology will allow the design of a more sustainable product.

7 CONCLUSION

A design tool using the Durham team-based methodology has been proposed and tested. Initial tests suggest that, in going beyond Lagerstedt's EFM, the DFSM may well be a substantial aid to the Redesign and New Product Design processes.
Further testing of the DFSM and follow-up to see whether or not it promotes follow-on work are required before any complete conclusions can be made but the initial indications are that this new methodology provides good information about the product and areas for redesign.

8 ACKNOWLEDGEMENTS

The authors would like to thank FKI Industrial Drives, Durham Pipeline Technology and Cambridge Consultants for their time and help in testing the DFSM.

References

1. Jeswiet, J. *Teaching mechanical engineering courses relating to the environment.* in *Design and Manufacture for Sustainable Development 2003.* 2003. Cambridge: Professional Engineering Publishing Limited.
2. Lagerstedt, J., *Functional and environmental factors in early phases of product development - Eco Functional Matrix*, in *Department of Machine Design.* 2003, Royal Institute of Technology: Stockholm. p. 156.
3. Stigson, B. *How Can the Corporate Sector Make an Effective Contribution to Sustainable Development?* in *World Business Council for Sustainable Development.* 2003. London.
4. Short, T.D., et al., *CIAM and North-Eastern Industry: The Road to Sustainability*, in *Design and Manufacture for Sustainable Development*, B. Hon, Editor. 2002, Professional Engineering Publishing Limited.
5. Datschefski, E., *The Total Beauty of Sustainable Products.* 2001: RotoVision.
6. United Nations Environment Programme, *Cleaner Production*, www.uneptie.org/pc.cp/understanding_cp.htm
7. Appleton, E. and J.A. Garside, *A Team-Based Design for Assembly Methodology.* Assembly Automation, 2000. **20**(2): p. 162-169.
8. Appleton, E. and J.A. Garside. *A Team-Based Design for Manufacture Methodology.* in *ASME 2000 Design Engineering Technical Conference.* 2000. Baltimore, Maryland, USA.
9. Department of Trade and Industry, *Manufacturing Best Practices*, www.dti.gov.uk/mbp/bpgt/dti-ma.pdf

MAAP – a method to assess the adaptability of products

B WILLEMS, J DUFLOU, and **W DEWULF**
Department of Mechanical Engineering, Katholieke Universiteit Leuven, Belgium
G SELIGER and **B BASDERE**
Institut für Werkzeugmaschinen und Fabrikbetrieb, Technical University Berlin, Germany

ABSTRACT

In order to achieve a sustainable growth, it is necessary to pursue the decoupling of economic welfare from the use of virgin resources: a shift from a linear economy to a closed loop economy is required. Consequently, adaptation processes such as maintenance, repair, upgrading and remanufacturing are needed. The adaptability of a product at the end-of-life is significantly influenced by its design. Therefore, a dedicated Methodology for Assessing the Adaptability of Products (MAAP) is presented in this paper. The method determines the improvement potential within the design of a product, its components and joints, using an adaptability metric. The MAAP is validated using a number of case studies.

1. INTRODUCTION

Current management practices and the standard of living are associated with an increase of resource consumption. The demand for production, use and disposal of products increases due to population growth and expanding requirements (1). Sustainable limits are exceeded and the available resources will be exploited in medium-term. The increasing demands can only be fulfilled in an ecologically compatible way, if energy and resource consumption per person are reduced drastically. The objective is to achieve more use with less recourses (2). Hence it follows, that the use-efficiency of recourses must increase. This increase can only be reached by a sustainable closed loop economy considering ecological as well as economical necessities. Legislation has become a key driver for the introduction of a closed loop economy. Several international approaches focus on avoiding negative environmental impacts that are related to products (3,4). They enforce environmentally friendly planning and design of future product generations, as well as regulations for the end-of-life treatment (5) and encompassing an extended producer responsibility (EPR) for present and future products. To implement the sustainable closed loop economy concept, products must be designed considering their entire life cycle, starting from the development, along their use, up to their reuse or disposal. More use can be obtained with the same resources if products are reused, with reuse being considered as the highest form of resource recovery. Prerequisite for product reuse is the feasibility of the required adaptation processes, such as maintenance, repair, remanufacturing, upgrading and downgrading. The adaptability of a product at the end of its life cycle is significantly influenced by its design. Thus, methodologies and metrics to assess and quantify a product's adaptability are required. This paper introduces a methodology referred to as the Method to Assess the Adaptability of Products (MAAP) that serves that purpose. This methodology supports the identification of improvement potential in the design of the product, its components and joints. The methodology is validated in a case study by applying it to a range of products.

2. THEORETICAL BACKGROUND

In literature, examples of efficient and effective use of metrics in product design evaluation can be found in the Design For Assembly (DFA) (6) and Design For Remanufacturability (DFR) methods (7). The DFA method, as developed and validated by Boothroyd and Dewhurst, owes its broad success to the simplicity of the method together with the fact that the metrics offer a good measurement for the assemblability of a product and a wide range of application possibilities. Despite the fact that the remanufacturability metric (7) also provides a good measure for the evaluated parameter, compared to the DFA approach, they experience more difficulties finding their way towards industry. The complexity of this method is much higher, designers have to invest more time and effort to perform the analysis and the required amount of data is significantly higher than is the case for the DFA procedure. Furthermore the remanufacturability metric is focussing on one option for the end-of-life treatment processes only. Since remanufacture is often considered not essential in the product life cycle (PLC), in contrast with assembly processes, the expected benefit of this method compared to the invested effort is often perceived as rather low.

Both methods are relevant but somewhat limited to one specific process. Due to environmental changes and legal consequences, like take-back legislation, designers need integrated tools that incorporate more options for the PLC. This is the objective of the newly developed MAAP described in this paper. The main difference between the proposed method and the product design evaluation methods referred to above is the scope of the procedure. First, the range is much wider: not only assembly or remanufacturing processes are evaluated, also maintenance, repair, up- and downgrading are assessed. Secondly the analysis goes more in depth. Where the other methods are only taking time criteria into account, the MAAP is looking to the whole list of product demands. This is necessary because not every requirement can be directly translated into a time parameter. Finally the profound nature of the method is a benefit. Apart from evaluating the product design as "good" or "bad", the method also offers a guide to locate improvement potential. In this way the designer is provided with an all-round analysis tool.

3. METHOD DESCRIPTION

3.1. Structure of the adaptation metric
As stated in Chapter 2, adaptation includes several processes. This assessment method considers the ones most frequently used: remanufacturing, maintenance, repair, upgrading and downgrading. The aim of the MAAP is to evaluate the suitability of a product to undergo those four processes. This will be represented by means of a metric, μAdaptation. The more

this value tends to 1, the better the design suits adaptation, the more it tends to 0, the worse the evaluation outcome. A secondary aim of the method is to locate potential improvement areas in the product design. Therefore the method strives towards a transparent structure, in a way that the user will be able to track the cause of low scores. The transparency is guarantied by using several submetrics, one for each of the four processes, which determine together the adaptation metric (Equation 1).

$$\mu Adaptation = \sum(\mu Remanufacturing, \mu Maintenance, \mu Repair, \mu Up/Downgrading) \qquad \textbf{(Equation 1)}$$

From literature [10] a list with all demands for the four processes was set up. Considering all requirements for each process, four categories could be defined to group the demands: the parts, connectors, spatial and time category. After linking those requirements, it becomes clear that almost all parts, connectors and spatial related criteria have a direct influence on the time criterion, which means that there is potential to eliminate them. However this would imply an elimination of the tracebility of causes. Since calculating the extra, non-time related metrics does not require any extra input of the user, this certainly proves its benefit.

It is obvious that certain demands for one process will be identically the same for the other three processes. This is mainly the case for the first three categories (e.g. having an ideal number of parts is a criterion valid for each process). For this reason these demands were isolated from the process specific requirements into generic ones. In this way seven submetrics can be identified: four process specific ones (μRemanufacturing', μMaintenance', μRepair' and μUpDowngrading') and three generic ones (μSpatial, μConnectors and μParts). Those form the metrics of an intermediate level. Since they still have a rather large scope, a basic level of metrics was created. Here each subcriterion evaluates just one specific parameter. The structure of the μAdaptation is graphically presented in Figure 1. The hierarchical structure and the large set of submetrics guarantee the tracebility of the causes of low metric values.

3.2. The first and second level metrics
Based on literature [6-10] all issues that were considered as potentially relevant for creating an adaptation metric were identified. Since the MAAP only aims at evaluating product design, the scope of the selected criteria was narrowed down to design driven issues.

3.2.1. Generic metrics.
As described in Section 3.1, there are some process specific requirements and some generic ones. The generic metrics include all demands concerning the composition of the product without considering the disassembly sequence. This means all requirements related to the parts, the connection of the parts and the spatial issues of those connections. These demands are respectively grouped into the parts metric, the connectors metric and the spatial metric.

Because of the proven utility of the DFA index of Boothroyd and Dewhurst (6), it was decided to base the generic submetrics on a similar concept. The use of an actual versus a theoretical minimal amount ratio for a specific evaluation parameter provides a good indicator of the suitability for adaptation. This formula can be applied to calculate all of the generic metrics (Equation 2).

$$\mu_{Parameter} = \left(\frac{\text{Ideal \# for a specific parameter}}{\text{Real \# for a specific parameter}} \right) \quad \textbf{(Equation 2)}$$

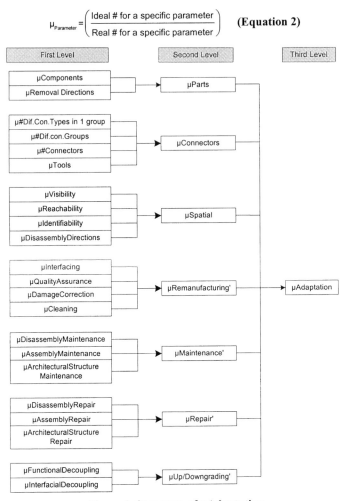

Figure 1. Structure of μAdaptation

Because many requirements use the theoretical minimal number of parts in their evaluation, the determination of this value is treated in detail here. According to Boothroyd and Dewhurst (6) parts must fulfill at least one of these three following criteria to be theoretically necessary:
- Large relative motions between the part and its adjacent parts.
- The part needs to be of different material then the other parts for functional reasons.
- The part is required to facilitate assembly and disassembly processes.

Bras and Hammond added an additional constraint for remanufacturing (7):
- The part isolates wear to parts with relatively low intrinsic value.

Because in the MAAP the scope is again broader, a supplementary condition can be added:
- The part has a different functional lifetime than the adjacent parts.

 D004/011/2004

This means that when a part breaks down, wears or needs maintenance more easily then adjacent parts, it should be able to be removed without affecting more durable parts. From the moment that the product satisfies one of these criteria, it can be considered as theoretically necessary.

The first group of generic metrics evaluate the number of parts and the number of removal directions per part, both parameters to be minimized. It is obvious that redundant parts or complex disassembly routes will decrease the adaptation efficiency because they increase the adaptation costs and slow down disassembly and assembly processes. Those two demands are grouped together into the μParts metric.

The second group of demands assesses the amount of connectors, the amount of different connectors groups, the range of various connectors within such a group and the tools used to (dis)assemble the connections. It is obvious that the disassembly process will speed up if the number of connectors is limited and in case several connectors are permitted, they are as similar as possible. Also for the associated tools the designer should strive to a large degree of standardisation. All these requirements are grouped into the μConnectors metric.

The last group of demands evaluates issues like visibility, reachability, identifiability of connectors and the disassembly directions on the product level. This last criterion reflects whether the product structure is platform based, in which case the product can stay in its initial clamping until the end of disassembly, or whether it is module based, which implies that the clamping of a product must change in order to facilitate the disassembly of certain components. All these elements are grouped in the μSpatial metric.

3.2.2. Remanufacturing.

Bras and Hammond already developed a profound analysis tool for remanufacturability (7)]. As remanufacturing is an integral part of adaptation, and the output of the analysis tool is also a metrical value, this Design For Remanufacturability index can be used in MAAP. The remanufacturing metric is composed of four categories for which a set of metrics have been developed:

1. Part interfacing, composed of disassembly and assembly metrics
2. Quality assurance, composed of testing and inspection metrics
3. Damage correction, composed of repair, refurbishment and replacement metrics
4. Cleaning

The development of the remanufacturability metric was based on Boothroyd and Dewhurst's DFA metric (Equation 3). However, apart from an assembly metric the DFR method also needed a disassembly metric. Following Bras and Hammond, the ideal disassembly time is estimated at 3s per part, unlike 1.5s for assembly processes. To know the efficiency of a disassembly process the ideal disassembly time versus the real disassembly time could be calculated (Equation 4). Because assembly and disassembly are activities required for each process of adaptation, both of these equations can also be used for the three remaining processes.

$$\mu_{Assembly} = \left(\frac{(\#Ideal\ parts) \times (1.5sec)}{(Time)} \right) \quad \textbf{(Equation 3)}$$

$$\mu_{Disassembly} = \left(\frac{(\#Ideal\ parts) \times (3.0\ sec)}{(Time)} \right) \quad \textbf{(Equation 4)}$$

For the explanation of the other first level metrics for remanufacturing, we refer to (7). As a final step in the analysis procedure, the assessments of all subprocesses are added in such a way that the *μRemanufacturing'* is obtained.

3.2.3. Maintenance and Repair.

Since maintenance and repair are processes with a lot of redundancy, they will be discussed together. Both processes first need partial disassembly until the level of the part to be maintained or the broken part is reached. Then the corrective actions, like maintenance or repair, take place and at the end the parts are reassembled again. Since the corrective action is very product specific and the time needed depends on the skills of the operator, they will not be discussed in detail. Instead all attention will be focussed on the product architecture, because this has substantial influence on the suitability of a product for maintenance and repair. The higher the probability that one specific part will need maintenance or repair, the earlier it should be located in the disassembly chain. This criterion is translated into μArchitecturalStructure (μAS). The failure information for maintenance or repair is provided by an FMEA analysis. Because it is unrealistic to require a high disassembly level for each part in the product structure that sometimes breaks down or needs maintenance, the problem will be narrowed down. For this reduction process the Pareto rule is applied. From maintenance management literature (11) it can be concluded that 20% of all causes is responsible for 80% of the effects. Based on this concept only the most important causes are taken into account when calculating *μArchitecturalStructureMaintenance*, μASM or *μArchitecturalStructureRepair*, μASR (Equation 5).

$$\mu_{ArchitecturalStructure} = \sum_{i=1}^{\sum effect>80\%} \left(\frac{B\%_i}{\sum_{i=1}^{\sum effect>80\%} B\%_i} \cdot \left(1 - \frac{DL_i - 1}{DL_{max}}\right) \right)$$

(Equation 5)

$B\%_i$ = Breakdown percentage of ith part

DL_i = Disassembly level of the ith part

DL_{max} = Maximum number of disassembly levels

$\sum effect<80\%$ = Summations untill the sum of the breakdown% of is >80%

i = Numbering for parts sorted according to decreasing breakdown percentage

As a consequence of the nature of these two processes, an operator will only need to disassemble until he has reached the targeted part. This will affect the disassembly time and subsequent reassembly time. To calculate the efficiency of disassembly until a certain part, the μDisassembly will be calculated from Equation 4. Because it concerns partial disassembly, this formula will be calculated separately for each part of the product. The time factor in Equation 4 represents the time needed to disassemble all parts on the path between the undisassembled product and the part for which the metric was calculated. Based on data from a Failure Mode and Effect Analysis the μDM (*μDisassemblyMaintenance*) or μDR (*μDisassemblyRepair*) are obtained by weighted adding of the separate disassembly efficiencies for all parts. The partial assembly efficiency (μAM, *μAssemblyMaintenance* or μAR, *μAssemblyRepair*) can be found in an analogous way by using the assembly Equation 3 and times instead.

The method for adding all first level metrics into the respective second level metrics *μMaintenance'* and *μRepair'* is described in Section 3.3.

3.2.4. Up/Downgrading.

Similar to maintenance and repair, upgrading and downgrading require also partial disassembly and reassembly processes. But apart from that, their demands are more related to modularity.

The architectural structure, disassembly and reassembly efficiencies (μASUD, μDUD and μAUD) are calculated like the partial disassembly efficiency for maintenance and repair. In this case the information of the FMEA will be replaced by data from the upgrading and downgrading planning.

Ulrich describes (10) two clear requirements for modularity: a product must have decoupled functions and decoupled interfacial relationships. The first parameters can be evaluated by counting the number of functions per part and the parts per function. This gives respectively μFunction/part and μPart/function. Together they will determine *μFunctionalDecoupling*.

$$\mu_{function/part} = \left(\frac{\#\,Parts}{\#Functions\ per\ part} \right) \quad \textbf{(Equation 6)}$$

$$\mu_{part/function} = \left(\frac{\#\,Functions}{\#Parts\ per\ function} \right) \quad \textbf{(Equation 7)}$$

The *μInterfacialDecoupling* is calculated by comparing the real with the ideal amount of decoupled interfaces.

$$\mu_{interfacial\ decoupling} = \left(\frac{Real\ \#\ of\ decoupled\ interfaces}{Ideal\ \#\ of\ decoupled\ interfaces} \right) \quad \textbf{(Equation 8)}$$

Finally, the assessments of all subprocesses are added, as described in Section 3.3., so *μUp/Downgrading'* is obtained.

3.3. The third level metric

Combining the second level metrics into a single adaptation metric is done by inversed weighted addition. This procedure ensures that the magnitude, idealization, annihilation and weighting criteria are fulfilled [9]. The same procedure is used for adding the first level metrics into the second level metrics.

$$\mu Adaptation = \frac{1}{\dfrac{w_{reman.}}{\mu Remanufacturing} + \dfrac{w_{main.}}{\mu Maintenance} + \dfrac{w_{repair}}{\mu Repair} + \dfrac{w_{up/down}}{\mu Up/Downgrading}} \quad \textbf{(Equation 9)}$$

$$\mu Adaptation = \frac{1}{GF + 0.5 \cdot \left(\dfrac{w_{Reman.}}{\mu_{Reman.'}} + \dfrac{w_{Main}}{\mu_{Main.'}} + \dfrac{w_{Repair}}{\mu_{Repair'}} + \dfrac{w_{Up/Down}}{\mu_{Up/Down'}} \right)} \quad \textbf{(Equation 10)}$$

with:

$$GF = \left(\frac{w_{spatial}}{\mu_{Spatial}} + \frac{w_{connectors}}{\mu_{Connectors}} + \frac{w_{parts}}{\mu_{Parts}} \right);$$

$$\mu_{Reman.} = \frac{\mu_{key\ repl.}}{\dfrac{w_{interfacing}}{\mu_{interfacing}} + \dfrac{w_{Q.Ass.}}{\mu_{Q.Ass.}} + \dfrac{w_{Correction}}{\mu_{Correction}} + \dfrac{w_{Clean}}{\mu_{Clean}}}; \quad \mu_{Main.} = \frac{1}{\dfrac{w_{DM}}{\mu_{DM}} + \dfrac{w_{AM}}{\mu_{AM}} + \dfrac{w_{ASM}}{\mu_{ASM}}};$$

$$w_{Reman.} + w_{Main} + w_{Repair} + w_{Up/Down} = 1; \quad \mu_{Repair} = \frac{1}{\dfrac{w_{DR}}{\mu_{DR}} + \dfrac{w_{AR}}{\mu_{AR}} + \dfrac{w_{ASR}}{\mu_{ASR}}}; \quad \mu_{Up/Down} = \frac{1}{\dfrac{w_{DUD}}{\mu_{DUD}} + \dfrac{w_{AUD}}{\mu_{AUD}} + \dfrac{w_{FD}}{\mu_{FD}} + \dfrac{w_{ID}}{\mu_{ID}}};$$

The weights of the individual submetrics from Equations 9 and 10 are determined by setting up a prioritisation matrix (Figure 2). The values are decided based on literature data and feedback from end-of-life workshops. The same procedure can be used to determine the other weights of the first level metrics. The user can change these relative weights in order to reflect the companies' product policy.

3.4. Practical use of the MAAP

Due to the broad scope of the method and the number of used metrics, it becomes time consuming to calculate every index by hand. Therefore, the MAAP was implemented as a spreadsheet. To remain the transparency of the procedure, each second level metric is calculated on a dedicated worksheet.

	Remanufacturing	Maintenance	Repair	Up/Downgrading	Total Score	Weight (%)
Remanufacturing	1	5	0.2	1	7.0	21%
Maintenance	0.2	1	0.1	1	2.0	6%
Repair	5	10	1	5	21.0	62%
Up/Downgrading	1	1	0.2	1	3.0	9%
					33.7	100

Legend

10 (row)	requires much more investment than (column)
5 (row)	requires more investment than (column)
1 (row)	requires the same investment than (column)
0.2 (row)	requires less investment than (column)
0.1 (row)	requires much less investment than (column)

Figure 2. Prioritisation matrix

At the end all values of the submetrics were put into Equation 10 to obtain µAdaptation. A graphical representation shows what categories are responsible for the magnitude of µAdaptation. Typical values for this metric are 30-70%. However the user should be aware that it is best to use these metric values in a relative way instead of as absolute numbers. This means in comparison with µ values of other similar products.

Based on the results of the metrics the tool also offers guidance how to improve the evaluated product towards adaptation. For every submetric with a value lower than 0.6 potential improvement actions are described with a reference to the relevant worksheets. Each worksheet then guides the user to those parts or connectors that have the largest potential to be improved. This whole procedure can be found in the software version of MAAP, available on request.

D004/011/2004

4. VALIDATION OF MAAP

4.1. Case Studies

To test whether the method is indeed effective to determine the adaptability of products, the MAAP procedure was validated by means of executing several case studies. Nine different product groups were selected from the Waste of Electric and Electrical Equipment category, including cellular phones, water cookers, video recorders, ironing machines, vacuum cleaners, washing machines, CRT monitors, coffee machines and personal computers. For each product group, several products were analysed, which made it possible to compare alternative product designs. These kinds of products are of particular interest, because they account for a large fraction of the total negative environmental impact caused by consumer products. Due to large production volumes and characteristically short time scales of technological and stylistic obsolescence, the impact of these products, associated with the production, use and end-of-life phase, is particularly high. Some landfilled or incinerated products, such as mobile phones, create the potential for release of heavy metals or halocarbon materials from batteries, printed wiring board, liquid crystal displays, plastic housings, wiring, etc. The fact that the components in these products are typically required to fit into a tight enclosing space even worsens this problem, which makes disassembly for adaptation a challenging task (12). The selection of the products for each product group was done in order to achieve a representative sample for the market situation in terms of age and brand diversity.

Before performing the assessments, all products where subject to some manual adaptation processes executed by specialised operators from profit and non-profit organisations. The products were disassembled, adjusted and reassembled again, in order to locate the weak spots in each product. After having this information, the assessment of each product could start. The results of this evaluation process are summarised in Table 1.

Table 1. Summary of the case study results

	μParts	μConnectors	μSpatial	μRemanufacturing	μMaintenance	μRepair	μUp/Downgrading	μAdaptation
Mobile phone 1	0,642	0,626	0,902	0,294	0,697	0,476	0,356	0,573
Mobile phone 2	0,651	0,705	0,896	0,238	0,611	0,525	0,307	0,576
Mobile phone 3	0,576	0,727	1,000	0,331	0,516	0,449	0,277	0,577
Mobile phone 4	0,625	0,804	0,949	0,238	0,684	0,526	0,267	0,588
Mobile phone 5	0,576	0,716	0,891	0,378	0,682	0,519	0,203	0,586
Mobile phone 6	0,454	0,563	0,760	0,219	0,508	0,448	0,290	0,477
Mobile phone 7	0,690	0,778	0,910	0,347	0,680	0,487	0,469	0,633
Personal Computer 1	0,535	0,550	0,739	0,423	0,544	0,555	0,631	0,574
Personal Computer 2	0,605	0,690	0,902	0,403	0,285	0,310	0,361	0,514
Water Cooker 1	0,682	0,670	0,752	0,401	0,444	0,318	0,498	0,524
Water Cooker 2	0,548	0,729	0,925	0,181	0,632	0,598	0,585	0,568
Coffee Machine 1	0,649	0,775	0,716	0,343	0,244	0,248	0,242	0,451
Coffee Machine 2	0,890	0,532	0,880	0,478	0,344	0,361	0,501	0,555
Iron Machine	0,620	0,904	0,827	0,281	0,228	0,292	0,344	0,494
Monitor	0,723	0,358	0,645	0,365	0,178	0,220	0,283	0,369
Vacuum Cleaner 1	0,518	0,814	0,973	0,368	0,561	0,280	0,491	0,514
Vacuum Cleaner 2	0,746	0,961	0,929	0,523	0,372	0,234	0,389	0,516
Video Recorder 1	0,698	0,556	0,835	0,250	0,285	0,271	0,331	0,448
Video Recorder 2	0,680	0,537	0,850	0,253	0,337	0,315	0,461	0,477
Video Recorder 3	0,592	0,548	0,945	0,205	0,308	0,292	0,321	0,444
Washing Machine	0,574	0,420	0,780	0,133	0,275	0,181	0,217	0,318

From the assessments it seems that *Phone 7* is best suited for adaptation, while the *Washing Machine* scored worst. This confirms the observations of the manual adaptation processes. The *Washing machine* contains a lot of connectors that are neither visible nor reachable and in some cases even rusted, in such a way that the disassembly process slows down significantly. Additionally these high time consuming connections are located at the beginning of the disassembly chain, which makes it worse, because in case of partial disassembly, this part will always be element of the critical disassembly path. Furthermore the disassembly process is complicated because the machine needs to be turned over repeatedly in order to be able to reach the desired parts. All these issues can be found in the assessment. *Phone 7* on the other hand, scores well for almost every metric. Only the remanufacturability and the repair metrics could be improved. This matches reality: the product has a logical product structure, not too many redundant parts and clear, simple connectors. Attaining a high µadaptation value for this product is consequently not surprising.

Apart from the numerical metrics that are obtained, the MAAP also indicates the weak spots in the product design with the highest impact on µAdaptation. To illustrate this, we focus on

D004/011/2004

the analysis of the *Personal Computer 1*. The individual scores for each submetric are given in Table 1. These values are presented in Figure 3.

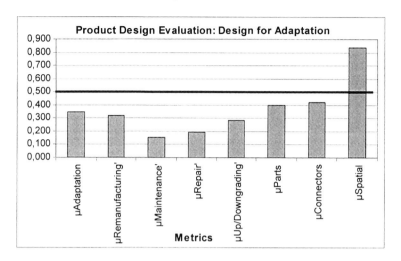

Figure 3. Graphical representation of the submetrics for the *Personal Computer 1*

From Figure 3 it is clear that the value of each metric, except µSpatial, is situated below the average of 0.5. This means that the evaluated Personal Computer is not suited for adaptation in comparison with its alternatives. MAAP indicates the following reasons for this unsuitability:

· µParts	Too many components
	➤ 69 instead of 29 (ideal case)
· µConnectors	Too many different types of screws
	➤ 7 types instead of 1 type
· µSpatial	Too many changes in disassembly directions
	➤ Disassembly should be executed in a single same direction instead of turning the product several times.
· µRemanufacturing	(Dis)Assembly time is too high
	➤ Many screws which require a lot of unscrewing time
· µMaintenance/Repair	Those components that need adjustment are situated very low in the product structure
	➤ The operator needs a lot of time to reach them
· µUp/Downgrading	Many product specific components and interfaces
	➤ Limited modularity, which makes upgrading difficult

Based on this evaluation the following suggestions for improvement of the *µAdaptation* score were made:
·   Reduce the use of screws
·   In case screws cannot be excluded: use the same type of screws

- Place the components that need a lot of adjustment high in the product structure
- Reduce the amount of abundant parts
- Adjust the product structure such that only limited disassembly direction changes are needed.
- Facilitate visibility and reachability of connectors.

These suggestions can be used as input for the redesign of the Personal Computer.

4.2. Preliminary correlation analysis of the results

The set of case studies, described in Section 4.1, are used to determine the effectiveness of the method. By means of a correlation analysis on the achieved results, it is possible to determine whether the developed metric is a good measure for adaptability. A parameter that can be used to determine the quality of the metric is the average time for adaptation per part (Tpp), calculated by dividing the total adaptation time by the number of parts. This variable was chosen because it can be directly translated into an economic value. Longer times will increase the adaptation costs and make the product less suitable for adjustment. The time value is calculated per part, to eliminate size differences between the products.

The correlation analysis between the calculated μAdaptation values and the Tpp of the respective products gives an indication of the quality of the metric. Based on the data of the case studies, a strong correlation of -0.91 was found between these two parameters (Figure 4). This value is high enough to assume that the MAAP result is a good indicator for the adaptability of products.

A second way to validate the method was to link the findings of the manual operators with the results of the MAAP analysis illustrated in section 5.1. After comparison of the data, it seemed that all problems that occurred during the execution of adaptation processes, were located by MAAP. Moreover 40% extra problems were indicated. These are problems where the operators are seldom confronted with during adaptation.

Based on these findings, we can conclude the MAAP is a valuable method to aid designers during the design phase of new products.

D004/011/2004

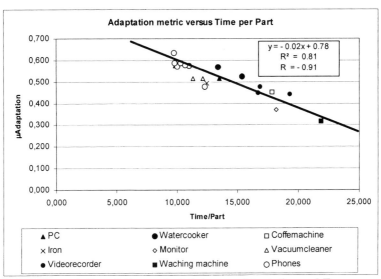

Figure 4. Correlation analysis between μAdaptation and Tpp

5. DISCUSSION

In this paper a metric-based Method to Assess the Adaptability of Product design (MAAP) was described. The approach is an elaboration of the widely used Design for Assembly and Design for Remanufacturability methods. The developed procedure offers designers the opportunity to evaluate their product design towards the four main processes of adaptation: remanufacturing, maintenance, repair, up and downgrading. The metrics were applied for several case studies selected from the WEEE category. Statistical analysis pointed out that the evaluation results correspond well with the findings from industry and that the μAdaptation metric is effective. The method has been implemented in a spreadsheet-based prototype in order to offer potential users a simple, semi-automated evaluation tool for assessing their products towards adaptation. The model calculates several submetrics that are added together in one final metric. Based on this information, potential improvements areas are pointed out. However, this method is subject to some limitations:

- The evaluation method has a subjective nature. (E.g. the user can decide on the weights of the different submetrics according to his own preferences) This limits the utilisation of the evaluation results: only comparing results obtained with the same user settings provides valid results.
- Due to the profound character of the method, the evaluation process requires a lot of data input and a lot of time to gather the information and complete the analysis. Naturally a detailed evaluation method is indisputably connected with a large amount of valuable output data. This is a trade-off the designer should make.
- The metrics can only be calculated during or after the embodiment phase of the product. This means from the moment that a satisfying level of information is available.
- The method only addresses problems in a product design stage. Issues like end-of-life policies are not taken into account.

Future research in this area will focus on:

Executing more case studies, in number and in product range, in order to be able to determine suitable generic weight factors.

Examine the effect of the uncertainty of the input data on the final result.

Improving the ergonomics of the spreadsheet by making the tool user-friendlier.

REFERENCES

1   **Seliger G., Müller K., Perlewitz P.,** More Use with fewer Resources, Proceedings of the 4th CIRP International Seminar on Life Cycle Engineering 1997, Berlin, pp 3-16.

2   **Seliger G., Basdere B., Keil T., Rebafka U.,** Innovative Processes and Tools for Disassembly, CIRP Annals 2002 Manufacturing Technology, San Sebastian, Spain, 2002, pp. 37-40.

3   **Directive 2002/95/EC** of the European Parliament and of the council, Directive on the restriction of the use of certain hazardous substances in electrical and electronic equipment (RoHS) of 27 January 2003.

4   Draft proposal of the European Council for a directive on establishing a framework for Eco-design of end use equipment of October 2, 2002.

5   **Directive 2002/96/EC** of the European Parliament and of the council on waste electrical and electronic equipment (WEEE), January 27, 2003.

6   **Boothroyd G., Dewhurst P.,** Design for Assembly: A Designers's Handbook, Boothroyd and Dewhurst Inc., Wakefield, Rhode Island, 1983.

7   **Hammond R., Bras B.A.,** Design for Remanufacturing Metrics, Proceeding of the 1st International Workshop on Reuse, S.D. Flapper and A.J. de Ron eds., Eindhoven, The Netherlands, 1996, pp. 5-12.

8   **VDI 2243**, Recycling-oriented Product Development, Beuth Verlag, Berlin, Germany, 2002.

9   **Ishii K.,** Modularity: A key concept in Product Life-cycle Engineering, Stanford University, Stanford, 1998.

10 **Ulrich K.,** The Role of Product Architecture in the Manufacturing Firm, Research Policy, Vol. 24, 1995, pp. 419-440.

11 **Villacourt M.,** Failure Modes and Effect Analysis (FMEA): A guide for continuous improvement for the semi-conductor equipment, International Semetech Inc., Austin, USA, September 30, 1992.

12 **Seliger G., Basdere B., Skerlos S., Morrow R., Chan K.,** Economic and Environmental Characteristics of Global Cellular Telephone Remanufacturing, Proceedings of the IEEE International Symposium on Electronics & Environment, Boston, USA, 2003, pp. 99-104.

Exploiting digital enterprise technology in sustainable design and manufacture

D BRAMALL, P BAGULEY, and **P MAROPOULOS**
School of Engineering, University of Durham, UK

ABSTRACT

There is a perceived dichotomy between the aim of increased product lifetimes and recyclability, promoted by sustainable development practitioners, and the reduced product lifecycle and cost-effective production methods demanded by the marketplace. To address these issues, this paper presents an assessment of the available digital enterprise technologies for the concurrent design of sustainable products and the associated manufacturing systems. Some of the tools and techniques which are available to increase the sustainability and manufacturability of products through innovative manufacturing methodologies are investigated.

1 INTRODUCTION

This paper presents a commentary on the use of a novel architecture for product and process design, developed jointly by the Universities of Durham and Oregon State, in sustainable design and manufacture. This technology is discussed in the context of the recent move towards agile manufacturing in distributed and collaborative manufacturing networks.

1.1 Digital Enterprise Technology for Product and Process Design

Digital Enterprise Technology (DET) is an architecture for product and process innovation [1] which is aimed at assessing and controlling the manufacturability of products at an early stage in their lifecycle, and allows for planning and synchronisation of work across the extended enterprise. DET is defined as 'the collection of systems and methods for the digital modelling of the global product development and realisation process, in the context of lifecycle management'. As such it represents an emerging paradigm for product lifecycle management in which the enterprise can take advantage of advances in product modelling, process planning, simulation, workflow management software and advanced laser metrology to develop and manage new, self-sustaining business processes.

DET has a 'heterarchical structure', with functional software components in one or more of the five main categories as shown in Figure 1. Each implementation of DET is unique and is configured through the flexible, internet based integration of data repositories, distributed systems and user sites. The internet and product data management systems are the backbone of DET, with standards for the communication and exchange of data being of primary importance. Typically, DET is introduced within an organisation to develop products faster and at reduced cost to cope with the reduced product lifecycles demanded by the marketplace. This paper shows how DET may also be considered as a framework for the technical analysis of engineering (and associated logistics) activities in which the environmental impacts of a product's design can also be addressed during the earliest stages of design. Convergence of product, process and resource models [2] and the emergence of digital enterprise standards mean that much of the manufacturing processes can be synthesised at varying stages of design according to the level of design data available.

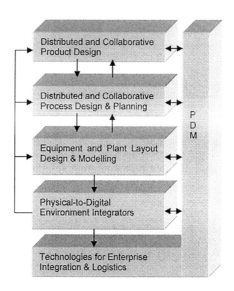

Figure 1 The DET Architecture

1.2 Production in Networks
DET is seen as fundamental to establishing future manufacturing strategies spanning global product development, production networks and agile manufacturing. Production networks [3] have the following characteristics;

- A number of cross-company partners involved in the manufacture of a product
- Mutual use of resources in the extended or virtual enterprise
- Ability to dynamically re-configure to take advantage of marketplace

The main reasons for forming production networks are, firstly, to take advantage of specialist skills and equipment and secondly, to allow for highly scalable production systems capable of reacting quickly to changing market conditions. Additionally, reason for carrying out distributed production in a network is to facilitate the local production of goods and services designed elsewhere, thus eliminating the need for expensive, time consuming and environmentally un-friendly packaging and transportation. These benefits arise through;

- Cooperative and distributed design and planning
- Capturing and making use of design and process knowledge with detailed data exchanges between customer and supplier
- Web-centric process planning using near real-time capacity and e-business systems

Enterprises that make use of a distributed production network face challenges related to planning and synchronising manufacturing operations, and estimating and controlling the downstream impacts of design decisions. Hence, DET plays a major part in controlling sustainability aspects of a product's manufacturing requirements as well as the quality, cost and manufacturing time of products.

2 EXPLOITING DET TO DEVELOP SUSTAINABLE PRODUCTS

The flexible arrangement of the software elements in DET means that product development and manufacturing processes can be designed to accommodate sustainable development issues alongside traditional manufacturing analyses such as Design for Assembly and Design for X [4]. Configuring a DET architecture to accommodate sustainable design and manufacture will primarily involve four factors; (1) optimisation of process, equipment and supply chain configuration, (2) data management, (3) mass customisation and (4) (re)manufacturing.

2.1 Optimisation in Early Design Decision Making

2.1.1 Multi-Criteria Optimisation for Process Planning
Sustainable design involves identifying how a product gives rise to environmental impacts during its lifecycle (raw materials, manufacture, distribution, use and end-of-life) and then investigating how to reduce these through optimising the design. Unfortunately, these sustainability criteria are numerous and often conflicting, hence for early product analysis it is critical to develop multi-criteria optimisation and evaluation techniques, especially when designing products that necessitate innovative manufacturing processes or utilise a collaborative approach to the production of assemblies. Because of the large number of alternative manufacturing options available in an extended enterprise, the use of evolutionary computing techniques and emergent synthesis approaches [5] is becoming more widespread for resolving these conflicting design objectives. Aggregate process planning [6] is an example of such a technique which has been developed to provide the automatic generation and technical evaluation of early process plans based upon quality, cost and delivery criteria (QCD), where the early technical analysis is carried out before designs are finalised and costs locked in. The semi-automatic planning procedure identifies the major manufacturing steps and assist the user in deciding upon the best allocation of components to factories in the network. This technique currently relies on a simulated annealing method and an objective function which selects near optimal solutions based upon cost. The quality and delivery performance indicators generated from the process plan are converted into a cost based upon loss functions.

The extension of these methods to give an early appreciation of costs related to environmental and sustainability factors would be technically feasible. The major changes required will be to the underlying data structures and process models. These changes would be necessary to store the data related to calculating a 'sustainability cost' based upon the specific candidate product, process and resource selected during the optimising routines and allow trade-offs between the conflicting requirements of lifecycle costs and environmental impact of disposable products. Essential cost-based data may relate to transportation, packaging costs, end-of-life disposal costs.

The requirements for the development of metrics of sustainability for use during early planning are;

- The system must cope with complex products and large scale resource models
- The system should balance environmental factors with existing business drivers
- A process plan must be generated without precise specifications or complete data

Technologies for the next generation of enterprise-wide DET systems are being researched, so that the early evaluation of process plans can be exploited in a production network to provide:

- "Resource-aware" planning for multiple sites
- Widening of the optimisation domain to include environmental and sustainability factors
- Identification of the critical path (shortest lead time) or shortest distance (best environmental performance) with respect to logistics and technology

2.1.2 Control of Manufacturing Waste using DET

A fundamental of sustainable development is the avoidance of the creation of 'waste'. This has many parallels with the engineering concept of continuous improvement of Kaizen [7] which drives organisations to eliminate operations that do not add value. Continuous improvement is an approach to increasing the effectiveness of manufacturing operations by eliminating waste. Four of the '7 wastes' [7] contribute directly to sustainable performance; overproduction, defects, inappropriate processing, excessive transportation. DET models can identify the 'physical' wastes using simulations early in the design cycle before processing options are committed to. Here DET is of primary importance because it enables designers to appreciate and avoid creating these wastes during design rather than relying on manufacturing engineers to remove these wastes later.

2.2 Product Data Management for Environmental Factors

If engineering designs are to be developed to use more sustainable technologies, then the initial focus must be on capturing and using environmental information generated throughout the product's lifecycle. Secondly, there is the requirement to be able to effectively re-use the captured information, in areas such as aggregate process planning, without the need to re-input data. The ideal way to do this is through existing Product Data Management (PDM) systems, which can be configured to capture and control the relevant data and workflow tasks. For example, a sustainable product design may require environmental impact assessments for the product itself, the consumables and energy used in manufacture for example. Obviously, these tasks must be done in the most efficient and timely manner, for example the calculation of energy usage for a process can only be carried out when the cycle time of the process is known, that is after a process plan has been generated. In this situation PDM can track the

D004/019/2004

requirement for a series of tasks to be carried out on a product basis and store the results of the analysis. This functionality is likely to be further enhanced through the increased use of agent-based techniques which pro-actively act to monitor the data within the PDM systems and alert users to tasks which can be done or are awaiting data.

A further utilisation of PDM within a DET architecture is related to the internet-based collection and dissemination of data and results. This is the key technology that facilitates the enterprise in that it allows the execution of tasks to take place wherever required. For example a detailed process model and cycle time may be calculated by the manufacturer and the results communicated via the PDM system so that everyone is able to calculate the total energy.

2.3 Mass Customisation

Mass customisation [8] has been proposed as a generic method for delivering products to the customer's exact specification. Customisation for such products is often done on existing production lines but is left to the last possible moment, however, another way of exploiting the DET paradigm is to move the customisation functions physically nearer to the end user. This obviously, has commercial advantages but also has implications for the requirements for packaging large, possible cumbersome components.

2.4 Re-manufacture and the Control of Manufacturing Waste

A final exploitation of DET for sustainable design is related to the concept of re-manufacture, particularly reverse engineering applications. Given that many modern engineering designs, especially in plant and machinery, are constructed in a modular fashion (as a result of the move towards distributed manufacture), it should be theoretically possible to upgrade the individual components rather that replace the entire product. A potential application of DET in this case is to reverse engineer the existing worn out components and manufacture replacement items to fit existing integration points.

The requirements for using DET to facilitate re-manufacture and the control of manufacturing waste related to maintenance and renewal of spare parts are;
- Laser metrology for measuring the existing part and its environment
- Simulation to check the new part can be installed easily
- Design for disassembly to simplify the re-manufacture process

2.5 Summary

This paper has presented four possible uses of DET for in sustainable design and manufacture, which are summarised in Table 1. These applications are not unique and there are undoubtedly many other ways in which this approach, linking of product design with downstream analysis of manufacturing operations, can be used to lower the total environmental cost of a product.

Table 1 Summary of DET Areas and Application in Sustainable Design and Manufacture

Area of DET	Application	Exploitation
Distributed and collaborative product design	Mass customisation and design for sustainable development	Design using less raw materials/more environmentally friendly materials Design for energy efficiency during use Design for disassembly, end of life Higher quality products with long lifespan Tailoring of final design elements at local manufacturing sites
Distributed and collaborative process design and planning	Optimisation of appropriate production strategies and resources to gain most environmental benefit	Reduction in in-process waste & pollution Selection of efficient (low energy use) resources
Equipment and plant layout and modelling optimisation		Reduce impact of transport/distribution
Physical-to-digital environment integrators	(Re-)manufacture	Simulation of assembly logistics Re-engineering of upgraded products using large-scale metrology
Enterprise integration and logistics	Increased communication	Closer integration between members of product development team to achieve benefits outlined

3 FUTURE PRODUCT AND PROCESS INNOVATION STRATEGY

DET tools are found in many industries, but it is no coincidence that the high value, high risk aerospace sector are the major users of this technology [9]. The automotive industry is also beginning to realise the benefits that DET can bring, but many other sectors have yet to realise the full potential of these methods as shown in Figure 2.

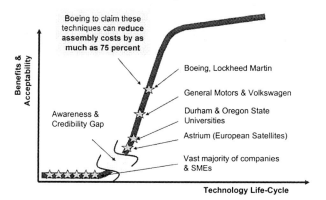

Figure 2 The DET Architecture

The current level of awareness is hampering the uptake of DET, particularly in small companies who do not have the time or capital to invest in training and development. However, increasing amounts of legislation relating to product lifecycle issues and customer demand for greener products should raise these issues up the agenda of small firms. The University of Durham is leading a range of technology transfer initiatives to address this situation giving practical project-based advice and assistance as well as training and education.

4 CONCLUSION

DET provides a framework for delivering product designs that are compatible with society's complex needs and address global challenges relating to production in networks and sustainability. Ultimately, cleaner design can save you money and increase profits by reducing waste, increasing product efficiency and encouraging innovative product development. DET can be applied to all stages of the design process, and sustainability considerations can be incorporated within the traditional design process. However, it is most effective when implemented at the start of the design process when the maximum number of design options are available.

In conclusion, as well as the generic industrial benefits of DET;

- Minimisation of risk in global product realisation
- Provision of analysis and computer support throughout the product's 'life cycle'
- Enabling 'digital manufacture and assembly' for complex products with short life cycles
- Integrated feedback can be received regarding; the status of production machines and plants, the capacity and logistics of the extended enterprise
- High plant reconfigurability, to meet product complexity and production network needs
- Low technology redundancy

There are also specific gains to be had by developing the technology to support the incorporation of sustainable development activities into the existing product development cycle;

- Mass customisation of products at local manufacturing sites
- Co-development of sustainable, environmentally-friendly production alternatives during product specification via multi-criteria design objectives; balancing environmental factors such as material selection, design for end of life with traditional quality, cost and delivery performance indicators
- Elimination of waste
- Re-engineering of complex products to maximise lifespan
- Increased awareness of environmental and sustainable factors during product development

REFERENCES

1 **Maropoulos, P. G.** (2003) A Novel Digital Enterprise Technology Framework for the Distributed Development and Validation of Complex Products. Annals of the CIRP, Vol 52, No. 1, pp 389-392.

2 **Brown, R. G.** (2000) Driving Digital Manufacturing to Reality. *In:* J. A. Jones, R . R . Barton, K. Kang and P. A. Fishwick, Proceedings of the 2000 Winter Simulation Conference, 10-13 December , Orlando, USA. 224-228.

3 **Wiendahl, H.-P. and Lutz, S.** (2002) Production in networks. Annals of the CIRP, 51(2), 573-586.

4 **Boothroyd, G., Dewhurst, P. and Knight, W.** (2002) *Product design for manufacture and assembly*. 2nd Ed. New York, USA: Marcel Dekker.

5 **Ueda, K., Markus, A., Monostori, L., Kals, H. J. J. and Arai, T.** (2001) Emergent synthesis methodologies for manufacturing. *Annals of the CIRP*, 50(2), 1-17.

6 **Maropoulos, P. G., Bramall, D. G. and McKay, K. R.** (2003) Assessing the manufacturability of early product designs using aggregate process models. *Proceedings of the Institution of Mechanical Engineers: Part B: Journal of Engineering Manufacture*, 217(B9), 1203-1214.

7 **Womack, J. P. and Jones, D. T.** (2003) *Lean Thinking*. 2nd Ed. London, UK: Simon and Schuster.

8 **Alford, D., Sackett, P. and Nelder, G.** (2000) Mass customisation - an automotive perspective. *International Journal of Production Economics*, 65(1), 99-110.

9 **CIMdata** (2003) *The Benefits of Digital Manufacturing: An Independent Report on Achieved Benefits and Return on Investiments with DELMIA Solutions*. Michigan, USA: CIMdata,

D004/019/2004

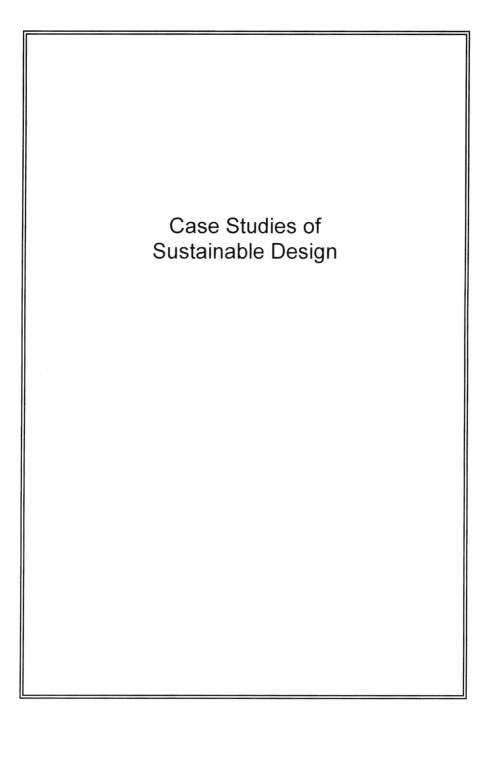

Case Studies of
Sustainable Design

Design decision-making and materials – investigations in the context of recycled polymers

E NORMAN, O PEDGLEY, and R COLES
Department of Design and Technology, Loughborough University, UK

ABSTRACT

Design decision-making can be viewed as the result of the interaction of knowledge, skills and values. The complexity of the design for sustainability agenda is explored in this paper through a discussion of the knowledge, skills and values associated with the designing of a polymer acoustic guitar. This was a case study for a recent PhD research project concerning the relationship of knowledge and designing. The role of values in design decision-making is being researched further through the analysis of designing with recycled polymers, and this paper:

- explains the choice of designing in recycled polymers as a context for exploring the role of values in design decision-making;
- outlines the research strategy being pursued;
- gives an indication of the results obtained from pilot studies and early research investigations.

1. DESIGN DECISIONS AND SUSTAINABILITY

It is widely acknowledged that the decisions designers make have a profound influence on sustainability issues. As designers determine the form and size, and select the material for a component, both the functional and environmental characteristics of a product are defined, as well, of course, as aspects that determine emotional responses. For example, the form, size and material choice determine the product weight directly, its cost, both as a result of the quantity of material used and the consequential manufacturing route, and the energy used for transportation and, potentially in use. It would be expected that optimum choices would tend towards 'minimisation' simply on the grounds of economic competitiveness. Why is it that all designers are not striving to do 'more with less' in the words of Buckminster Fuller? If economic rationality were not sufficient grounds alone, then the growing importance of environmental issues in designing might have been expected to further tip the balance in the same direction. Reducing material consumption results directly in less resource use in primary material extraction and the associated decrease in the loss of biodiversity and, of course, resource conservation for future generations. In relation to these kind of issues should not good design and sustainable design be synonymous?

If a range of acknowledged good designs are explored, then surely there should be an expectation of finding aspects of Datschefski's 'hidden beauty' within them (1). Is it the pursuit of 'total beauty' that drives designers? Or do other factors such as perceived market and human requirements swamp such idealism (rationalism)? The intention of this paper is to illustrate the complexity of this agenda through the analysis of the authors' designing, not to scrutinise the efforts of others. That is for them to do and their consciences to prompt.

2.DESIGN DECISION-MAKING AND THE POLYMER ACOUSTIC GUITAR

Design decision-making is essentially concerned with the outcome of the interaction of knowledge, skills and values (2, 3). A PhD research project carried out from 1995-99 by Dr Owain Pedgley and supervised by Dr Eddie Norman explored the role of knowledge in design decision-making (4). The research used the design and development of a polymer acoustic guitar as a case study, through which evidence was obtained from a diary of designing. A detailed discussion of the design of this PhD research programme can be found in Norman *et al* (5). However, in relation to the pursuit of total beauty, the following issues were some of those we wished to address.

- In the mid-1990s, low cost beginners' instruments were often poorly set-up and difficult to play, limiting participation. Surely modern manufacturing methods could deal with this.
- Some 250,000 unplayable, toy plastic guitars were being made and landfilled annually. Surely such 'design for landfill' could be put right. (Our guitars would be polymer and not plastic!)
- The manufacture of high quality, wooden instruments depends on the use of rare tropical tonewoods. Surely a contribution could be made to the preservation of such non-renewable materials for future generations.

Prior to starting the PhD research programme, a preliminary study had been carried out to establish that engineering science had little to offer to the designing. Polymers are essentially viewed by engineers as sound absorbers, not generators. So the research was concerned with the evolution of innovative technology and the transition of newly acquired knowledge from the tacit to the articulate (6). Interesting topics, but not as interesting as the 'blind prejudice' with which many greeted the project! Fortunately for the project an internationally renowned luthier, Rob Armstong, was supportive from the very first discussions.

Figure 1 Rob Armstrong with Guitar No.673 (of which Eddie Norman is the proud owner)

Figure.2 The certificate of ownership for No.673

Figure 1 shows Rob with Eddie's 'Armstrong acoustic' (No.673) and Figure 2 its certificate of ownership. Each of the 700 or so guitars that Rob has built is different. Each is a reflection of true craftsmanship and the intimate understanding of how individual wooden components vibrate in unison to create the guitar sound, which is the embodiment of the instrument's beauty. Rob always knew that a polymer guitar could work, and it seemed to him to be an entirely sensible aspiration. Mario Maccarferri (one the world's most famous guitar makers) had made polymer guitars in the 1950s and only really failed to achieve market success as a result of the polymers available at the time. Figure 3 shows Maccaferri's design, and he always remained a believer in the potential for polymer instruments. In contrast, the general opinion was that polymer instruments could never sound as good as wooden ones, and would always be perceived as low value artefacts … without people actually seeing or hearing one!

Figure.3 The polymer acoustic guitar designed by Maccaferri (7)

Figure.4 The SWD wood guitar made by Martin (1)

Some of the more knowledgeable critics point to the approach taken by Martin in developing the 'SWD wood guitar' (shown in Figure 4). Datschefski reports:

> More than 70 per cent of the wood used in its construction is harvested from forests independently certified by the Rainforest Alliances 'SmartWood' programme and Scientific Certification Systems, both of which operate in accordance with the rules of the industry's governing body, The Forest Stewardship Council …

An interior label indicates that a portion of the proceeds from the sale of the guitar will be donated to the Rainforest Foundation International. The RFI was founded to support indigenous peoples and traditional populations of the rainforests in their efforts to protect their environment and their rights.' (1)

At first sight this sounds fair enough and might be fine, if Martin were able to build such guitars in a sustainable way and at a price that all new guitarists could afford. Until you become aware that the majority of the guitars in shops in Western countries are built in only a small number of factories in the Far East and later 'branded', it sounds quite realistic. The 'out of factory price' of such guitars is typically about £20 (in 2004) and it is impossible for Western makers to compete. It is only mid-range and high-quality guitars that can be economically made in the West: producing sustainable, wooden beginners guitars for the main Western markets would depend on the continued employment of 'low cost labour' in the Far East. Although the approach embodied by the Martin SWD model might address many environmental issues, the economic and social aspects of sustainability need more thought.

3. THE EMOTIONAL DIMENSION OF DECISION-MAKING

So, given that we can actually make them, what is wrong with polymer guitars? (Any doubters of the sound quality that can be achieved might consider listening to Gordon Giltrap playing and endorsing our guitars on the cool acoustics website, 8). The answer, of course, is that people's responses are governed by their emotional responses to materials, as much as by rationality. The economic value of a product, and values in general, are cultural constructs and designers are educated to, and practice within, the same constructs. Despite the scientific research behind the development of polymers, and despite the technological achievements in extracting oil from ever more challenging places, people still do not value polymers. They are liberally discarded and current UK recycling rates indicate how much we care about where the next tonne of polymer comes from. So, research into designing in polymers offers a rich context for the exploration of the role that values play in design decision-making. (In the same way that the polymer acoustic guitar offered a rich context for the exploration of knowledge and designing).

Products made from recycled polymer materials can invoke even more polarised responses (9). Figures 5 and 6 show two products designed by Industrial Design and Technology undergraduates at Loughborough University – John Vann and James Duder.

Figure.5 John Vann's electric violin made from recycled polymer (HIPS)

Figure.6 James Duder's 'love chair' made from recycled polymer (HDPE)

John Vann decided that it was necessary to spray paint his polymer violin, but James Duder decided that this was not necessary for the 'love chair'. The violin was m ade from HIPS (high impact polystyrene) and was part of an on-going investigation concerning the hidden beauty of compression formed recycled materials. (These came from Smile Plastics Ltd, 10). Some of the emerging knowledge from the polymer guitar project concerned the key influence of 'holes' on the vibration behaviour of polymer sheets. Foamed polycarbonates are the basis of the patent which emerged from the project (11). It was hoped that the voids in compression formed recycled sheets might exhibit similar properties (this is still part of an on-going investigation). Such would be their hidden beauty. However, John Vann judged that the market was not ready for such a complex judgement and that spray painting would give the necessary 'surface beauty'. The outdoor environment of the love chair is however already essentially variegated and James Duder was surely right in judging that the surface finish with its marble-like qualities was appropriate as it was. However, designs from such recycled materials can now achieve a level of controversy, which was not perhaps possible when the materials were designed and developed by Jane Atfield at the Royal College of Art in the 1990s. In 1996 her RCP2 chair (Figure 7) was included in the *Recycling: Forms for the Next Century - Austerity for Posterity* exhibition that toured the UK. In introducing her work she wrote as follows

Figure.7 Jane Atfield's RCP2 chair

'While at the RCA (1990-1992) I reused leftover materials f rom f actory p rocesses s uch a s i ndustrial recycled felt for armchairs. I also incorporated found objects into my work and began importing recycled plastic from America. This led to a two year project researching and developing a similar post-consumer recycled plastic material, made from high density polythene from empty shampoo, milk or detergent bottles such as Domestos or Frisk, which I now sell through the company Made of Waste. My motivation with this material has been to respond to environmental issues and to extend the use of discarded objects into a new and evocative material. I aim to promote the acceptance of recycled materials as viable alternatives to virgin materials and to extend their application beyond one-off and batch produced items. Therefore, other considerations including functional requirements, cost factors and the quality of the object are paramount'. (12)

4. EXPLORING THE ROLE OF VALUES

So, when it came to designing the PhD programme to explore the role of values in design decision-making to be started by Rhoda Coles in 2002, designing in recycled polymers seemed to be the ideal context. The most significant previous attempt to differentiate and categorise values in relation to designing was made on behalf of the Assessment of Performance Unit (APU). In their publication concerning *Understanding design and technology,* they used four categories

- Technical values (e.g., flexibility, precision and confidence);
- Economic values (e.g., value, price and cost);
- Aesthetic values (e.g., self expression, workmanship and proportion);
- Moral values (e.g., impact on the environment, religion and needs) (2).

In 1993 Professor Phil Roberts noted an additional dimension of values

- Hedonic values (e.g., the senses, desires and demands) (13).

This area was no doubt considered in 1982 (private communication from Professor Phil Roberts), but was not included in the final report at the time. This could be interpreted as reflecting a change in culture. Starting in the mid 1980's (14) and especially by the early 1990s (15), the significance of emotional factors in design decision-making was becoming ever more recognised. Even more recently studies have highlighted the link between a material and these less tangible values, the 'associations it carries, the way it is perceived and the emotions it generates' (16).

It becomes clear when we look at certain products that Roberts hedonic factors and tacit emotional responses have a huge impact within design. If we compare a polystyrene cup to one made of glass, they are visually nearly identical (Figures 8 and 9), and it is not until we engage our tactile and acoustic senses that we notice a difference. In much the same manner

it is the sensory pleasantness, or hedonic value achieved when a guitar is played that gives its appeal.
It is also clear to see an understanding of these more emotional values is present within industry, in fact in some forward thinking companies such as

Figures 8 and 9. Polystyrene and real glass cups (19 and 20)

Phillips (17), research into social, environmental and economic values has become an important part of R&D.

'It is still widely assumed that technology drives growth. However history shows us that technological innovation is a strategically important condition but, if it is offered to a society that is culturally, socially, and economically unprepared to accept it, its value will be lost...' (18).

Figure 10 Phillips 'Living memory' project used an understanding of social

Figure 11 The 'Q4 plugged' was developed through a realisation that the values

It is also clear that some companies are using these emotional responses to over design elements within their products to reflect the values that are embedded within them, 'The use of Allen screws to mount the machines steel filler cap of the Audi TT and the prominent welds of the mountain bike express the engineered robustness of both products' (16)

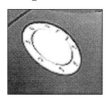

Figure 12 The filler cap, along with other small parts of the Audi TT is engineered to look robust in order to reflect the values embedded in the design of the rest of the vehicle. (20).

Nevertheless, the pilot studies (21) soon demonstrated some of the 'grey areas' within our current understanding of the use of values in design decision-making. Through a thorough literature review (22), brainstorming and seminar discussions RC has now developed an appropriate categorisation system for use in this research project (Table 1) and the proposed recording system (Table 2 where NDD stands for novel design decisions, see 23). RC is now proceeding towards the main data gathering activities and hopes to have time to investigate both the influence of age (e.g., school and university students and professional designers) and training (e.g., across the industrial and engineering design spectrum) on the role that values play.

Table 1. Rhoda Coles categorisation system for values

External values	Internal values
Societal values	Perceived societal values
Identified stakeholder values	Perceived identified stakeholder values
Economic system values	Perceived economic system values
Values embedded in design	Designer's personal values
	Meta-values

Table 2: Rhoda Coles' categories for values influencing the design decision-making process

Decisions / Values	External Values				Internal Values				
	Societal Values	Identified Stakeholder Values	Economic System Values	Values Embedded in Design	Perceived Societal Values	Pr'd Identified Stakeholder Values	Pr'd Economic System Values	Designer's Personal Values	Meta-Values
Decision 1									
Decision 2					NDD				
Decision 3								NDD	
Decision 4		NDD							
Decision 5									
Decision 6									
Decision 7									
Decision 8									
Decision 9									
Decision 10									
etc								NDD	
TOTAL									
		NDD							
					NDD				
								NDD	
								NDD	

The main data gathering activities will take place over two projects; the first will be a day in which participants will be asked to complete the re-design of a lectern mainly using recycled polymer sheet available from Smile Plastics (10), or from a specified recycled polymer of their own design (such as the one used for the lectern in Fig 13 which uses aluminium logos moulded into recycled polycarbonate sheet); the second will be a week long project to produce concepts for electric guitars, also predominantly using recycled plastic sheet that will present data over a more longitudinal project. Before the commencement of the projects information regarding participant backgrounds and expertise will be sought as well as their preferences for a selection of materials (these will then be related to the project outcomes in qualitative discussion). During the design activity participants will have access to a number of resources to allow the project work to be as realistic as possible.

Data will be collected using two methods; a retrospective interview (see 21,22) at the end of the day/week will provide information regarding design-decisions over the complete project; and a short concurrent verbalisation and protocol analysis (*ibid*) taken during the projects will give a more complete perspective of design-decisions taken during a small section of the design task.

Figure 13. The development of a lectern designed by RC

In order to discuss the relationship between the use of values in decision-making and the success of eco-design outcomes, the final designs from the two projects will be produced using CAD (to ensure they are all offered in the same style) and presented to relevant consumers. Favoured designs can then be discussed in relation to the design-decisions that produced them.

5. CONCLUDING DISCUSSION

It is becoming evident that design decision-making is driven by a very complex agenda. The complexity of this agenda needs to be researched and properly understood, particularly in the context of furthering the cause of sustainable design. Figure 14 shows the design of the polymer acoustic guitar which was exhibited at the Frankfurt Musikmesse in March 2002. Because of the low cost and comparatively easy modification possibilities, the backs for the prototypes were made of foamed polyurethane by reaction injection moulding. As the only polymer material which gives outstanding sound quality, the guitar soundboard was made of extruded foamed polycarbonate. The bridge was made from

Figure 14. The prototype polymer acoustic guitars for the Frankfurt Musikmesse (8)

recycled acrylic and a different recycled polymer was used for the top nut and saddle (tradename: Tusq). Components were bonded with a combination of methylmethacrylate and epoxy adhesives. So what happened to the recognised 'rules of thumb' of sustainable design, e.g., avoiding the use of too many different polymers and avoiding bonding dissimilar polymers. These are generally reasonable principles, but are essentially concerned with disassembly, and are guitars intended to be disassembled?

Curiously, and unthinkingly, it has been an acknowledged goal of the project for some years to move towards a welded polycarbonate (PC) design. That way, it would be possible to destroy the guitar, shred it and recycle it without any contamination. Technically this is becoming ever more possible, but from a sustainability perspective, is it the right direction to go? Would the consequences of the additional environmental pollution resulting from the increased use of PC instead of other polymers outweigh the potential advantages at a (possible) end of life? A full lifecycle analysis might one day provide an answer, but in the meantime, the agenda will inevitably be driven by the designers' values (heuristics, rules of thumb) and skills (e.g., pattern recognition). The essential point of this paper is that asking designers to design sustainably is 'asking a lot' and the decision-making support tools available need to be a great deal more sophisticated before there can be any realistic chance of success.

References

1. Datschefski E (2001) *The total beauty of sustainable products*, Rotovision

2. Hicks G (1982) *Understanding design and technology*, Assessment of Performance Unit

3. Norman E (1998) 'The nature of technology for Design', *International Journal of Technology and Design Education*, 8, 67-87

4. Pedgley O F *Industrial designers' attention to materials and manufacturing processes: analyses at macroscopic and microscopic levels*, PhD thesis, Loughborough University, 1999

5. Norman E, Heath R and Pedgley O (2000), 'The framing of a practice-based PhD in design, *http://www.core77.com/research/thesisresearch.html*, 1-14

6. Harrison, G (2002) *The continuum of design education for engineering*, The Engineering Council, London

7. Gruhn G and Carter W (1993) *Acoustic guitars and other fretted instruments: a photographic history*, Miller Freeman Books: San Francisco

8. Cool Acoustics website, www.cool-acoustics.co.uk

9. Norman E (2001) 'Creating markets through designing with recycled polymers', *Proceedings of first international conference on (eco)design for profit: achieving commercial success* , Environmental Business Network: Yorkshire and Humber, University of Sheffield

10. Smile Plastics recycled plastic sheet, www.smile-plastics.co.uk, accessed 05/05/04

11. Norman E, Pedgley O and Armstrong R (1999), *Acoustic device (patent)* Patent Number GB99 19922.6

12. Atfield J (1996), *Recycling: forms for the next century - austerity for posterity,* Craftspace Touring, Birmingham

13. Roberts P (1993) *The Royal College of Art schools technology project – project paper – the purposes of design and technology in education*, Royal College of Art, London

14. Goonatilake S (1984) *Aborted discovery: Science and creativity in the third world*, Zed books, London

15. Layton D (1992) *Values and design and technology – Design curriculum matters: 2*, Department of Design and Technology, Loughborough University of Technology

16. Ashby and Johnson (2003) *Materials and design: the art and science of material selection in product design*, Butterworth Heinemann, Oxford

17. Phillips design website, www.design.phillips.com, accessed 05/05/04

18. Green J (2003) 'The value of valueing people' in *New Value News*, No 15, Phillips Design, January 2003, 21-23

19. WK Thomas website, http://www.wkthomas.com

20. Laithwaites website, www.laithwaites.co.uk

21. Audi filler cap, www.matey-matey.com/tt.shtml, accessed 05/05/04

22. Coles R (2003a) 'An exploration of the role values play in design decision-making and how this affects ecodesign outcomes' in B Hon (ed) *Design and manufacture for sustainable development*, Professional Engineering Publishing Ltd, London, UK, 117-132

23. Coles R (2003b) 'An exploration of the role values play in design decision-making' in J R Dakars and M J de Vries (eds) *PATT13 (Pupils' Attitudes to Technology),* University of Glasgow, 211-219

24. Akin O and Lin C (1995) 'Design protocol data and novel design decisions' in *Design Studies*, Vol 16, 211-236

Reduction and reuse of waste – design and new-generation plastics

C A CATANIA
Student at Department of Design, University of Palermo, Italy
F P LA MANTIA
Dipartimento di Ingegneria Chimica dei Processi e dei Materiali, Italy

ABSTRACT

Today, it is important to knowledge all the practicable ways to reduce and reuse waste.
The solution of this problem must include a general strategy about utilisation of the various recycling techniques, and/or energy recovery. Indeed, we are encouraged to do so, due to the large amount of packaging and disposable items. These products should be removed from the waste stream, reprocessed and used again.

In this work we will analyze systems and consumption model respectful with the environment, eco-efficient manufacturing processes, new generations of plastics (biodegradable plastics, green composites materials and biocomposites), and we will propose some new products can be manufactured using these new generations of materials with less environmental impact.

1 INTRODUCTION

Nowadays for sustainable development and to prevent degradation of the environment it is necessary to explore the fate of waste.
This is necessary if we want to reduce the amount of waste generated and to change the idea for which the society identifies materials as wastes.
We must minimise production of waste. The solution of this problem must include a general strategy about utilisation of the various recycling techniques, and/or energy recovery.

For this goal is important to acquire information to quantify the environmental advantage and to define criteria for the management of wastes as: waste prevention, source reduction, recycling and composting.
Moreover it is important to minimise the generation of waste through eco-efficient manufacturing processes.
Indeed, today, Design for Environment (DFE) in the phase of product development and innovation, assumes a strategic role: Design for Environment encompasses Design for Disassembly and Design for Recycling and is a new approach based on developing new materials and new design to reduce the environmental impact.

In this work the design of new desk accessories will be used in order to define smaller environmental impact . Then we will describe the product made in new generation plastic, as an example of application of these guide lines.

2. TO PREVENT WASTE

Municipal Solid Waste (MSW) is an important and high priority tool in the 21 century. The quantity of solid waste going to landfill since 1960 is continued growing. Indeed, material generated in 2001 are estimated 229 million of tons as shown in tab.1

Municipal solid waste (MSW) consists of items as: container, and packaging (soft drink bottles and cardboard boxes), durable goods (furniture and appliances), nondurable goods (newspaper, trash bags, and clothing), other wastes (food scraps and yard trimmings).

We can control this waste by reducing, reusing and recycling it.
Waste reduction efforts save money, energy and natural resources. Thus is important to make waste reduction a part of our life styles.

Table 1 Generation of material in MSW ,2001 (In millions of tons)

Source: Franklin Associates, Ltd

	Weight Generated
Paper and paperboard	81.9
Glass	12.6
Metals	
Steel	13.5
Aluminum	3.2
Other nonferrous metals*	1.4
Total metals	18.1
Plastics	25.4
Rubber and leather	6.5
Textiles	9.8
Wood	13.2
Other materials	4.2
Total Materials in Products	171.5
Other wastes	
Food, other**	26.2
Yard trimmings	28.0
Miscellaneous inorganic wastes	3.5
Total Other Wastes	57.7
TOTAL MUNICIPAL SOLID WASTE	229.2

Includes waste from residential, commercial, and institutiona
* Includes lead from lead-acid batteries.
** Includes recovery of other MSW organics for composting.

Reduction and reuse are the first priorities in the solid waste management hierarchy introduced by Environmental Protection Agency (E.P.A.).(1)
The components of the hierarchy are :

- Source reduction
- Reuse
- Recycling of material
- composting
- Combustion/incineration
- Landfilling

Environmental benefits of waste reduction (preventing waste, reusing , and recycling) are:

- Prevent pollution created by manufacturing new products
- Save energy in manufacturing, transportation and disposal of products
- Decrease greenhouse gases emissions, which contribute to global climate change
- Conserve natural resources such as, timber, water metals and fossil
- Reduce the need for landfill and incineration

Therefore we can to reduce the waste through a mix of practice:
Source   reduction: to reduce the amount of waste generated by changing the design , manufacture, purchase, o use of materials or products as:
Product design with less materials
Redesigning packaging to excess material
Includes purchasing durable, or be used again after its original use is completed
Source reduction prevents emission of many greenhouse gases, reduce pollutants, save energy, conserves resources, and reduces the need for new landfills.

Reuse: u sing a product more than once, for the same purpose or for a different purpose. Reusing, when possible, is preferable to recycling because the item does not need to be reprocessed before.

Recycling: the materials (items as paper, glass, plastic and metals) are collected and processed and manufactured sold and bought as new products. Prevents the emission of many greenhouse gases and water pollutants, save energy, Stimulates the development of greener technologies, conserves resources as timber, water, minerals, and reduces the need for new landfills. MSW recovered for recycling is shown in tab.2

Recycled products: are made from materials already discarded. Items in this category are made totally or partially from material destined for disposal or recovered from industrial actives like aluminum cans or newspaper.

Composting: Another form of recycling is composting that is the controlled   biological decomposition of organic matter, counting a high portion of biodegradable material and includes green waste as food and yard wastes.
The result of this decomposition process is "the composty" a crumbly, soil like material
So composting can reduce the amount of waste that ends up in landfills or incinerators.

Combustion: combustion with energy recovery is often called waste-to-energy , combustion without energy recovery is called incineration.

Landfills: are areas where waste is placed to the land, landfills present significant health and environmental risks if not well designed and maintained

Table 2 Generation and Recovery of products in MSW by material
(In millions of tons and percent of generation of each product)

Source: Franklin Associates, Ltd

	Weight Generated	Weight Recovered	Recovery as a Percent of Generation
Durable Goods			
Steel	10.9	3.0	27.8%
Aluminum	1.0	Neg.	Neg.
Other non-ferrous metals*	1.4	0.9	64.8%
Total metals	13.3	4.0	29.6%
Glass	1.7	Neg.	Neg.
Plastics	8.0	0.3	3.9%
Rubber and leather	5.6	1.1	20.1%
Wood	5.0	Neg.	Neg.
Textiles	2.9	0.3	11.8%
Other materials	1.2	0.9	73.7%
Total durable goods	37.6	6.6	17.5%
Nondurable Goods			
Paper and paperboard	43.5	15.6	35.9%
Plastics	6.1	Neg.	Neg.
Rubber and leather	0.9	Neg.	Neg.
Textiles	6.7	1.1	16.1%
Other materials	3.2	Neg.	Neg.
Total nondurable goods	60.4	16.7	27.7%
Containers and Packaging			
Steel	2.6	1.5	58.8%
Aluminum	2.0	0.8	40.0%
Total metals	4.6	2.3	50.8%
Glass	10.9	2.4	22.0%
Paper and paperboard	38.4	21.1	55.0%
Plastics	11.2	1.1	9.6%
Wood	8.2	1.3	15.2%
Other materials	0.2	Neg.	Neg.
Total containers and packaging	73.5	28.1	38.3%
Other wastes			
Food, other**	26.2	0.7	2.8%
Yard trimmings	28.0	15.8	56.5%
Miscellaneous inorganic wastes	3.5	Neg.	Neg.
Total Other Wastes	57.7	16.5	28.7%
TOTAL MUNICIPAL SOLID WASTE	229.2	68.0	29.7%

Includes waste from residential, commercial, and institutional sources.

*   Includes lead from lead-acid batteries.

** Includes recovery of other MSW organics for composting.

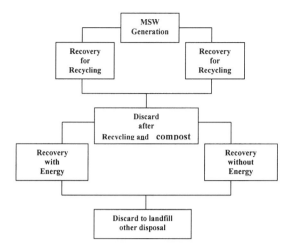

Figure 1 material flows methodology for estimating discards of products and materials in municipal solid waste

3 ECO-EFFICIENT MANUFACTURING FOR WASTE MINIMISATION

It is important minimizing the generation of waste through eco-efficient manufacturing processes. Thus Designers to find solution more environment friendly, will use Design for Environmental (DFE)

DFE is a design methodology that aiming at minimize environmental impact of products examing all aspects of the production of items including source of raw materials, manufacturing, distribution, use and disposal.

Within DFE, recyclability has received great attention among design and manufacturing engineers together with Design for Disassembly

Thus strategies for product waste minimisation are:
- Use of new material as biodegradable materials to replace non biodegradable materials
- Making items of recyclable materials
- Reducing total material content of packaging
- Developing reusable, long- life containers

A tool that can be used for these aims is Life Cycle Assessment that examines the life of the products from raw materials until final disposal.

The goal of LCA is to identify the environmental impact resulting from a product, process or industrial activity throughout its life cycle from the extraction of raw materials up to use and disposalfig.2

For each phase, the inputs (raw materials, resources and energy) and outputs (emissions into air and water, generation of solid wastes) are first quantified (LCI) and then generally added together to produce environmental impact indicators.

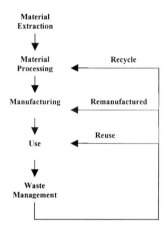

Figure 2 Scheme of life cycle

LCA include some benefits as:

Product improvement: LCA can identify the most efficient and cost effective options for reducing the environmental impact of a product or services

Process improvement : LCA can be used to assess a company operations and production processes. It is a useful way o quantify resources and energy use .

Strategic planning : LCA can be used for strategic planning As environmental regulations and expectations increase to improve their environmental operation

The LCA process can be organised in four phases:

Goal and scope definition: description of the goal, functional unit and system boundary, definition of what to include and what to exclude; **Inventory analysis**: collection of data for each step in the life cycle identification and quantitative evaluation of the materials and emissions to air and water ; **Impact assessment** : evaluation of the environmental impact, selected impact categories and modeling of the inputs and outputs for each category; **Interpretation**: review of all of the stages in the LCA and about the consistency of all the assumption

4. INTEGRATION OF MATERIALS WITH WASTE MANAGEMENT RECYCLING, COMPOSTING AND GREEN MATERIALS

Environmental emergency imposes on the current system of production new approaches for saving resources reducing the amount of toxic materials and managing wastes. Therefore if we want to reduce the volume of materials used is important to focus on materials management and new materials design

Objects, would be designed from raw material to ultimate disposability. For instance, the conventional plastic usage for utensils and packaging, single use disposable has created serious environmental problem because these short life items are not usually recycled.

A possible solution for these items is the used biodegradable plastic or recycled materials.

Biodegradable materials would satisfy on ecological and environmentally approach for product design use and disposal

When products are made using recovered rather than virgin materials, raw materials and less energy are used during manufacturing and fewer pollutant are emitted.

Recycling materials reduces the need to use raw materials, reducing primary process waste and air and water generated by these processes. Indeed recycling helps reduce or eliminate the pollution associated with the first two stages of material processing as show in fig.3

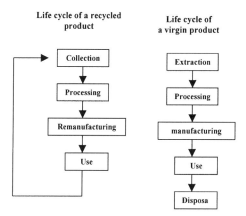

Figure 3 Less energy is needed to manufacture products from recovered materials than from virgin materials

4.1 Recycling

Recycling gives new opportunities to post-consumer material to be used in the some other application. Glass, metals, paper, can be easily recycled and part of plastic can be also recycled.

The technologies used for recycling depending on the material type: crushing and re-melting glass, shredding and pulping waste paper, shredding and re-melting steel, granulated and extruding waste plastic.

The recyclability of materials depends on technical and practical recyclability, (if it is easy or difficult to recover) and reprocess the material and the availability of a infrastructure for recovery and reprocessing

Recycling of metal have the highest market value, instead plastic recycling is not well development

The process to recover the steel is shredding and involve some steps:

toxic and hazardous substances are removed, appliances are put through the shredder, then the steel is sent to mills for reprocessing, the remaining materials, plastic, rubber, glass and wood (as residual) are sent to landfill.

Materials plastics are more difficult to recycle because: they need to be separated before recycling , have high costs of collection and processing, and low quality during the recycling process.

The benefit of Recycling

- Recycling saves energy
- Recycling reduces the need for landfill and incineration
- Recycling prevents pollution caused by the manufacturing of products from virgin materials
- Recycling helps sustain the environment for future generation
- Recycling conserves natural resources
- Recycling decreases emissions of greenhouse gases that contribute to global climate change
- Using recovered materials also generates less solid waste.

4.2 Recycling companies

The materials collected in recycling are valuable commodities and represent essential component of marketplace, in 1995 the US recovered 56 million tons of materials. today, companies that use recovered materials are a vital sector of the economy (2).

These Companies have adopted concept of Design for recycling.

Design for recycling means that materials used for the product could have a secondary life moreover for this goal is important to ensure that design for disassembly and recycling objectives are enclosed in the product and for this is important:

- Easing the removal parts
- Eliminating non reyclable or hazard use materials
- Reducing the variety of plastic used
- Using recycled and recyclable materials
- Avoiding parts that combine incompatible materials, such as plastics and metals

Car manufacturer have been among first to developed disassembly plants to recover materials (plastic, glass, metal) that are considered waste in the shredding process.

Today other companies as Whirlpool are active in promoting recycling. They have a program called Design to Build Equity in the End -of-Life Appliances. These industries are reducing the number of materials used, recover refrigerants and remove components. Another example are industries of electronic goods, they have a program to build new machines using recycled parts from used machines returned after leasing

4.3 Economic development of recycling

Recycling add Value to the Economy, creates Job and reduces demand landfill sites.
Recycling represents a significant force in the U.S. economy and makes a vital contribution to job creation and economic development (3).
Recycling creates business for collect, process, and recovered materials ,as the industries that manufacture and distribute products made with recovered materials
For example recycling study for the North Carolina, documented that recycling activities support 8,800 jobs in the state.
A study of 10 Northeaste states , recycling employed 103,00 people, 25 % in materials processing and 75 % in manufacturing. This show that some communities are realizing benefit from recycling

4.4 Composting

Is the biological decomposition of organic materials, such as paper, food wastes, yard trimmings, and wood by microorganism to produce compost of rich soil useful for improving the fertility of the rich.

Thus composting provides a convenient way to reduce the volume of trash and household products. It also provides a valuable product that can enhance the growth and health of the yard.

Thus in an integrated waste management system may be necessary use, recycle, and dispose of biopolymer material

4.5 Green materials

Today plastics are important in every aspect of our lives. They are an integral part of this century appearing in all types of products from packaging to advanced aerospace materials
Polymer composites are being used in place of more conventional metals . Indeed advanced composites made using fibres as graphite, kevlar, glass and are used in application from helicopters to tennis rackets and automobile parts.
Composite materials are not biodegradable or recyclable, and disposal of these composites is difficult indeed that they have become a larger part of the municipal solid waste (MSW)stream

Indeed Plastics are a rapidly growing segment of the MSW stream and the more important application for plastic are:
- containers and packaging (eg. soft drink bottles, shampoo bottles, lids)
- durable goods (appliances, furniture)
- nondurable goods (diapers, cups and utensils, trash bags)
- plastics in automobiles

Thus it is important an alternative to petroleum resources, an alternative that can replace materials derived from mineral base which renewable materials
Therefore there is a growing urgency to develop green materials and innovative technologies that can reduce other materials
For example an emerging as an alternative to glass fibre reinforced composites is natural fibres flax, hemp, kenaf as reinforced plastics.

Composites fabricated using these natural fibres are lighter, less expensive, partly biodegradable and are available as waste of agricultural resources in many countries. These materials used in sectors like automobiles and buildings.

In this work , agricultural fibre are examined as a reinforcing material to design and manufacture composites for product design development.
Natural fibres are renewable ,cheaper, biodegradable and can provide a solution to environmental pollution by finding new uses for waste materials. The mechanical properties of these material filled plastics are comparable to those of glass fibres.
Natural fibres can be classified as:
Seed fibres (cotton, kaprok), Bast fibres (flax, hemp, jute, kenaf,), Hard fibres (sisal, yucca, pineapple), Fruit fibres (coconut), Wood fibres
Natural flax fibre reinforced composites are better than syntetic fiber reinforced composites in properties such a biodegradability, ligh in weight, non corrosive, reduced environmental pollution and are recyclable.
Natural fibre reinforced composites, Green composites, from a new class of materials which seem to have good potential in the future as substitute for wood based material in many application. When the p lastic m atrix i s a b iodegradable p olymer t he composites a re c alled "biocomposites"

4.6 Green composites (nonbiodegradable plastic matrix)
They are composites combine nondegradable resins with degradable fibre.
Degradable fibre (flax, hemp, jute, kenaf, etc.) are used for reinforcing nondegradable thermoplastic polymers as polypropylene (PP), polyethylene (PE), polyvinylcloride (PVC), are used as alternative reinforcements to glass fiber in composites. These fibbers obtained from plant stems or leaves, are renewable annually. Compared to glass, these fibres provides better insulation against noise and heat in applications as automotive door panels separation.

Other composites made with wood fiber between 30% and 70% are used in window and door frames, outdoor decking, automotive panels and furniture. These plastic lumbers are made with recycled plastics and filler used are obtained from waste from sawmills, wood flour, waste wood products as packaging, pallet, construction wood scraps, are therefore inexpensive a nd a re a partial s olution t o w aste d isposal p roblem. O ne c omposite c alled medium density fibre (MDF) substitute boards for wood is obtained by resin/binders as polymeric diphenylmethane disocyanate (MDI) with wood fibre and flour

4.7 Biocomposites (biodegradable plastic matrix)
Organic fillers have been used also as manufactured objects with biodegradable matrices as Mater Bi, Polylactical polymers (PLA), etc., these materials are polymer based on natural sources like starch, sugar, etc. and when these biopolymers are reinforced with natural fibres, which are also biodegradable, these composites can be considered biocomposites.
Indeed at the end of their life they can be easily disposed or composted without environmental impact, fig.4.

 D004/012/2004

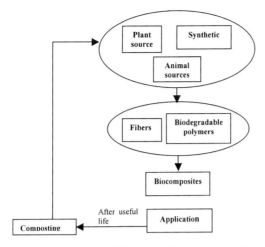

Figure 4 Life cycle of green composites

These composites may be used in many application as :
- •Nondurable products with short life cycles of one to two years
- •Products for short time (few time) use before disposal
- •Indoor application for several year, like wood.

These new composites have good properties and could replace plastic parts in the interiors cars and in packaging materials and other consumers products and they have excellent insulation against noise, but they are not as strong as Kevlar and graphite, they are low in cost, biodegradable and are renewable

5 MARKET OF COMPOSITES MATERIALS

Applications in some Companies
The market for such composites has grown and shows considerable growth in the future.
Indeed some companies as Can Fibre Group in Canada and the US produces many variety of MDF under the name of AllGreen. In India WBSL produce fibre boards, a waste products after sugar cane juice extraction. Flexform Technologies in US blends Kenaf and hemp fibres with PP and PET, fibre to produce composites for application, including automobile door panels. Phenix company in US has created with biocomposite technology a line of products can be used in interior application, these new materials used are:

Environ biocomposites
This i s a material made from recycled paper products, soybeans and colour additives and rapidly renewable agricultural resources. It can be used where the appearance of granite and wood is desirable. Environ can be used for commercial and residential furniture

Dakota Burl composites is a biobased material made from renewable agricultural resource it is ideal for a variety of interior applications.

Biofiber decorative Wheat composite

This biocomposite is created from a renewable agricultural resource and offers an alternative to traditional wood or panel products. It can be used in a variety of decorative interior applications.

6. CASE STUDY

In this work we show the development of ideas for sustainable products through the utilization of new and emerging green composites materials.

These materials have shown, in preliminary work, properties good enough for a variety of application for building, furniture and in some automotive sectors. They could be used as substitute for wood or in many application currently made from petroleum based plastic.

For this goal we have focused on the products for office that considerable contribute to waste stream. There are many different type of products used in offices: but beside office equipments and major consumables paper, toner cartridges etc., there is a small number of other office products, that currently contribute to waste stream. Among them desk accessories have been chosen for using these green composites as raw materials

6.1 Desk accessories in market

In market we have examined desk accessories

These products, designed to keep office organized and workspace more productive are:
card holder, pencil cups to file storage and clipboards, are available in various colours and materials as glass ,wood , metal, plastic.

Each piece is often sold individually.

6.2 Design of Desk organizer

In desk organizer, fig.5 the main characteristic is the presence of two elements, a base and a simple board, that without glue and interlocking systems (lives, rivets) constituted a single element . The board with simple holes can contains pen, pencil, card name, letter,etc.

The advantages of this desk organizer are:

It utilizes vertical space to maximize work space, it contributing to an improved environmental impact ,indeed, in this item is used a new material, as alternative to wood , plastic. This material is derived from organic fillers made with recycled plastics and its packaging is thought to be realized in a biodegradable material.

Thus through the redesign of the existing products, is possible to achieve a significant reduction of the resources and low environmental impact, therefore can be a partial solution to waste disposal problem.

The prototype in green composites fig.6 is made with recycled polyethylene from green houses, filled with 30% (by weight) sawdust (1mm diameter), a processed by twin screw corotating extruder.

The recycled polymer contains LDPE (65%-75%), LL DPE (10%-15%) and EVA (12%), plus some impurities. Sawdust comes from sawmill waste and was deparated by a mesh grid.

The extruder (D = 19mm and L/D = 35) was running at 200 RPM, with a thermal profile of 120-130-140-150-160-170-180°C.

6.3 Environmental requirements

Environmental requirements of the product are in table3

  D004/012/2004

The Design for Environmental process for this project is shown in fig.6

Table 3 Environmental requirement of the product

Manufacture	Use	Disposal
Use recyclable materials	Min. packaging	Easy disassembly (no glues, no rivets)
	Min. waste	
Biodegradable materials		Components recyclable
Renewable materials		

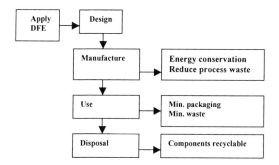

Figure 5 DFE for the life cycle of desk organizer

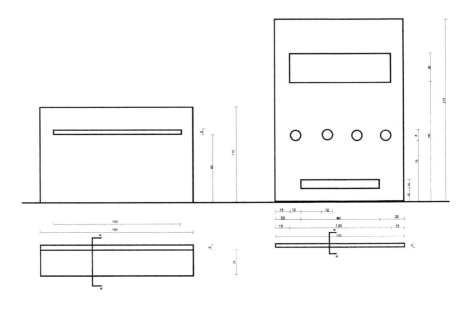

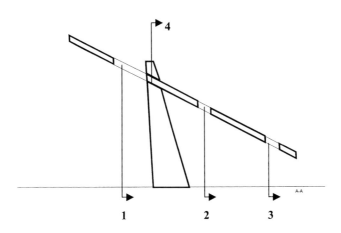

Figure 6 Desk organizer

1. letter holder
2. pen /pencil holder
3. card holder
4. union between two elements(without glue , lives, rivets)

D004/012/2004

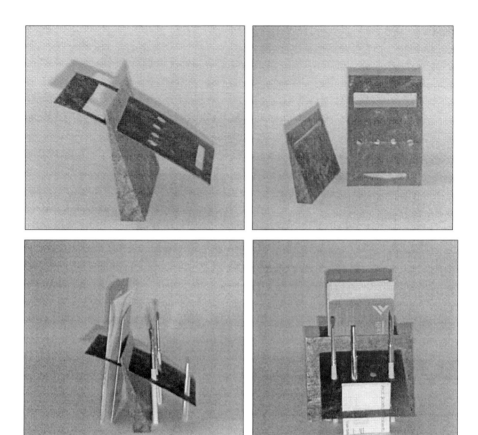

Figure 7 Prototype in Green composites

7. CONCLUSION

Plastic composites based on natural fibres have attractive properties, as light weight and are less expensive. Some natural fibres possess also good mechanical properties.
Preliminary work has shown their potential for applications in the building and automotive industries. It is important that these materials can have a new approach involving designer, engineers, environmentalist, to include their potential and their low environmental impact.

In this work an example is shown that suggest a new application of these green composites used as alternative to existing materials, as a possible solution to waste disposal problem.

REFERENCES

1 **EPA** (2003) Municipal Solid Waste in The United States: 2001 facts and Figures. EPA530- R-03-011.Washington, DC www.epa.gov

2 **EPA** (1997) Characterization of municipal solid waste in the United States:1996 update, EPA530-R-97-015. Washington, DC www.epa.gov

3 **Shore M. J.** (1995) The impact of recycling on jobs in North Carolina. Prepared for the North Carolina Department of Environment, Health, and Natural Resources, Office of Waste Reduction, Raleigh, NC

4 **La Mantia F.P., Morreale M., Mohd Isak Z. A** "Processing and Mechanical Properties of Organic Fillers-Polypropylene Composites" send for publication on Journal of Applied Polymer Science

5 **Lewis H.** and **Gertsakis J.** (2001) Design+ Environment. Greenleaf Publishing, UK

6 **Okubo K,** and **Fuji T** (2002) International Workshop on Green Composites, Japan

7 **Loddha, P., and Netravalli, A. N.**, (2000) Recent Advances in Polymer and Composites, Allied Publisher, New Delhi, India

8 **Brezet J. C.,** and **Van Hemel C. G.** (1997) Ecodesign: A Promising Approach to Sustainable Production and Consumption, Paris

9 **Beccali, G., cellura, M.,and Mistretta, M.,** (2001) Managing Municipal solid Waste. International Journal of Life Cycle Assessment,243-249

10 **Phenix** web site http:// www.phenixbiocmposites.com/products

11 **Bastioli, C.,** (1995) Biodegradable Materials and the separate/composting of organic waste. Environmental Conference and Exhibition on the Role of Biodegradable Materials in Waste Management" , Tokyo

Solar-powered bus shelter case study – design and development toward sustainability

B TUCKER, H MILLWARD, and **D MURPHY**
The National Centre for Product Design and Development Research (PDR), University of Wales Institute Cardiff, UK
R JAMES
BSW System Limited, Llandow, UK

ABSTRACT

This research study is based on a knowledge-transfer collaboration between The National Centre for Product Design and Development Research (PDR) and BSW Systems Ltd. The aim of the two-year collaboration has been to implement sustainable design procedures for the development and manufacture of a broad range of street furniture items. The research employs a case-study methodology, and focuses on the drive to minimise the environmental impact of a specific bus-shelter design by employing solar arrays to power internal lighting and information displays. The use of renewable energy would appear to bring obvious tangible benefits; however, a more detailed analysis of materials usage, battery-weight compensation and life-cycle servicing is required in order to estimate the true environmental impact of this product.

1 INTRODUCTION

New product development activities within traditional manufacturing companies provide higher financial returns than practically any other type of similar investment (1). However, companies within the UK continue to operate within a harsh economic climate with increased low-cost competition from overseas, and Government strategy advocates that manufacturing companies must move up the value-added chain in order to generate a clear competitive advantage. One mechanism by which to enhance and add value to product design is to incorporate ecodesign into the new product development process. It is widely acknowledged that companies must reduce the environmental impacts resulting from their activities. In addition to minimising this environmental life-cycle burden, effective ecodesign can also bring economic and social benefits to business and key stakeholders: specifically, greater resource efficiency, reduced hazards, increased control for reuse/remanufacture/recycling and added functionality (2). Johansson (3) has identified six factors that affect the successful integration of ecodesign: management, customer relations, supplier relations, development process, competence and motivation. It is recognised that many of these factors relate directly to the success of new product development in general, implying that the adoption of ecodesign procedures should complement and enhance the whole design and development process.

Comprehensive ecodesign activities have tended to be confined to larger industrial companies, who may be motivated to preserve the company image as well as the environment. The literature examining ecodesign within small and medium-sized enterprises (SMEs) is more limited, and these types of company are not sufficiently aware of the potential benefits of ecodesign (4). SMEs are often in a prime position to identify innovative environmentally-friendly products as a consequence of their close working relationships with customers and suppliers, but it takes time and effort to implement and maintain effective ecodesign procedures within this sector of the economy. It has been estimated that 95% of the three million businesses in the UK employ fewer than 20 people (5); consequently the performance of SMEs as product developers is a matter of no small concern, and a number of guidelines have been formulated to improve eco-innovation and product development of such companies (6). There can be a tendency for small companies to conduct product development in an *ad hoc* manner (7); therefore appropriate ecodesign procedures have the potential to refine this process and drive an overall improvement in costs, quality and lead-times within the SME sector.

This paper reports the initial ecodesign work undertaken by a small manufacturing company, based in South Wales. In response to new European markets and domestic environmental policy (Local Agenda 21), the company has started to develop comprehensive ecodesign procedures to be applied from concept development in the design office through to waste minimisation on the factory floor. The design work in this case study focuses on the early-stage development of a solar-powered shelter. A wide variety of products and services employ renewable energy, and the literature describing the application of solar-power covers many benefits (8). This research study reports on the rapid product development of a full production prototype and the long-term aspirations for a sustainable product. This two-stage approach to design is not 'right first time' but more akin to the 'prototype to learn' philosophy espoused by Kelly (9), and has helped to ensure that customer and environmental benefits are fed through to the next design iteration. The first production prototype, delivered early to a key customer, has highlighted a number of issues. For example: the numerous solar array options, improved structural integrity due to increased battery/roof weight, and the fixture implications of operating without national grid restrictions. For the transition to a more sustainable product, further market research was undertaken and Life Cycle Thinking employed, paying particular attention to materials selection and maintenance/servicing requirements. This paper reviews the initial results from this two-stage product design and development process, reports on the practical challenges associated with solar-powered products, and highlights key issues for SMEs adopting environmental considerations within a 'traditional' manufacturing scenario.

2 CASE-STUDY METHODOLOGY

BSW Systems Limited was established in 1979 and is now the largest private employer in the Vale of Glamorgan, South Wales, with a staff of approximately 80. It is a leading company in the design and manufacture of quality street furniture items. The range of products includes bus and rail stations, bus and rail waiting shelters, public seating, public information display boards, and other bespoke engineering projects for private and public organisations. These products are designed and developed in-house and supplied to order. The majority of the manufacturing and assembly operations are also undertaken in-house, within the company's dedicated aerospace-style production area. The company has approximately a 30% share of the UK shelters market and has established strong relationships with key customers, such as Network Rail. BSW have ambitious plans to increase their UK and European market share, and in partnership with PDR, have identified the need to introduce a 3D CAD-based 'design for the environment' capability to add value to their ongoing new product development projects and improve efficiency within their core manufacturing business. A collaborative TCS (Teaching Company Scheme) programme was established in February 2003, and this paper reports the preliminary results during the first 12 months of the programme.

PDR have employed the TCS model as an effective mechanism for partnership and collaboration with a wide range of SMEs, predominantly in Wales. TCS has been in operation for over 20 years, and is a government-backed knowledge transfer scheme. The aim of the scheme is to strengthen the competitiveness and wealth creation of the UK by stimulating innovation in industry through structured collaborations with universities and research organisations. TCS is run for the government by Technology Transfer and Innovation Limited (tti Ltd.). A typical TCS programme is a two-year partnership between one company, one university and tti Ltd. Each individual TCS programme is designed to address the key elements central to the successful development of the specific company. The two-year project provides employment for a well-qualified graduate TCS Associate for the duration of the programme. It should be noted that the TCS structure has been re-branded as Knowledge Transfer Partnership (KTP) for programmes starting after September 2003.

All the PDR-based TCS programmes are, or have been, focused on product design, and the numbers reflect the UK trend in that the majority have been based in SMEs. PDR have successfully completed 12 TCS programmes since 1995, seven of which have been with small companies. A typical PDR-based TCS programme implements a new design capability within a 'traditional' manufacturing company. In line with other researchers (10), we have found that the TCS model is an ideal vehicle through which to analyse the design-to-manufacturing interface and the associated elements that impact upon new product development within SMEs.

The well-defined management and structure of the TCS process promotes a detailed analysis of the company from the university partner's perspective. A close working relationship is developed with the company in the early stages during the drafting of the TCS grant proposal. This is written by the university in collaboration with the company, describing the company and the financial benefits, and quantifying the aims and objectives of the programme. A key feature of the proposal is a detailed 104-week Gantt chart, which defines a programme of work to address the strategic needs of the company. Following the approval process, a grant is awarded and PDR employs an Associate to work full time at the company for two years to meet the project objectives. The Associate is assigned at least one PDR-based supervisor and at least one supervisor from the company. The regular contact with the company fosters a level of trust and co-operation that generates an in-depth understanding of the subtle issues and problems inherent in any small company.

TCS programmes are characterised by a commitment to disciplined effective project management through mandatory monthly and quarterly meetings. The monthly meetings between the supervisors and the Associate not only focus on the technical issues within the programme, but also address training and personal development requirements. The quarterly meetings are designated Local Management Committee (LMC) meetings, and are underpinned by support from a tti consultant. LMCs act as the programme's steering group to ensure that the longer-term objectives for the company and Associate are met. The documentation (technical reports, executive summaries and presentation material) arising from the structured TCS meetings, in parallel with the weekly informal contact with the company personnel, results in a comprehensive portfolio of case study material, and this provides the foundation to this research study.

3 SOLAR-POWERED SHELTER DEVELOPMENT

3.1 Design brief

Cardiff Harbour Authority operates the Cardiff Barrage development, and has a commitment to environmental considerations through an integrated transport policy and the use of renewable energy sources. Consequently, the customer derived a specification for a solar-powered bus shelter for BSW to design and manufacture, the critical factors being:

a) Quality and Image
b) Sustainability
c) Functionality and 'Fitness for Purpose'
d) Compatibility with wheelchair and pushchair access
e) Effective protection from weather
f) Security and ease of operation
g) Ease of maintenance
h) Resistance to vandalism

Sustainability was high on the list, and this translates into employing a suitable form of photovoltaic technology to provide steady-state and on-demand power requirements. The elements requiring electrical power can be divided between (a) internal lighting, and (b) information panels for arrival/departure information. The system would need to operate in all periods of winter and summer darkness between 6.00am and midnight, and ensure five days' autonomy from any solar charging during the winter period. Furthermore, the shelter had to fit into a modern streetscape setting, and the operational life, maintenance, servicing and disposal life-cycle needed to be taken into consideration during the design and development stages. The styling and construction also needed to be appropriate for this type of emerging future market. The design of this shelter has clear implications for other projects that need to operate without reliance upon permanent connection to the national grid (e.g. rural developments).

3.2 Materials analysis

Materials selection, utilisation and processing play an important role when considering the overall environmental impact of a product; therefore, prior to the concept development of the solar-power shelter, a detailed analysis of the materials within a conventional BSW shelter was undertaken. The aim was to review the raw materials selection, manufacturing processes, assembly procedures and associated costs in order to provide the basis for a refinement of the design-to-manufacture process (e.g. through value analysis and/or life cycle assessment (LCA)). The shelter under consideration was a standard 2m, four bay model, produced using in-house batch manufacturing techniques. A fully detailed materials and manufacturing analysis is beyond the scope of this preliminary study, but the core materials can be summarised as follows:

- Mild Steel: predominantly used for roof beams, end gables, spigots and seating supports. Recent developments have also employed stainless steel.

- Aluminium: used in glazing supports, guttering and seating platforms.
- Glass-Reinforced Plastic (GRP): used to produce the roofing sections.
- Glazing: toughened safety glass or polycarbonate used as windows.
- Fittings: mechanical and electrical elements in a range of metals and polymers.

The manufacturing fabrication processes (machining, manual/laser cutting, bending and powder coating) are applied mainly to the mild steel and aluminium parts and assemblies. The GRP fabrication is sub-contracted. As a first approximation, the environmental impact of a standard shelter can be related to material weight, and the economic impact can be summarised in terms of raw materials' costs and manufacturing costs. Clearly the in-house fabrication processes also impact upon the environment, therefore excess manufacturing costs could correlate with environmental impact. A summary of material weights and the associated material/manufacturing costs is given in Figure 1. This shows typical relationships, for example the GRP contributes minimum weight but is responsible for a significant share of the costs. The main finding is that the aluminium components contribute significant weight and significant costs, therefore this is a prime area to minimise with regard to environmental and economic impact.

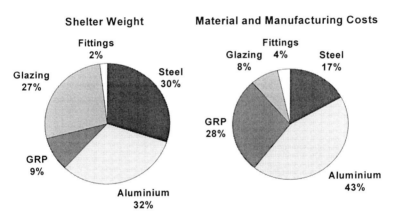

Figure 1. Summary of material weight and associated costs for a standard shelter design

3.3 Life cycle thinking
The materials analysis has provided a route towards a full materials and manufacturing inventory, which can form the basis of a conceptual LCA; however, these LCA tools require thorough expertise. The approach taken during the first year of the TCS programme has been to lay the groundwork for a conceptual or simplified LCA, and implement this when staff training and competence has progressed. In order to generate life-cycle appreciation, it was thought more practical to use published LCA results to provide the early-stage guidance. This approach is better suited to SMEs where time and cost factors preclude the use of specialised tools in favour of workable solutions such as a checklist or product design specification (PDS).

An LCA based on CO_2 equivalent emissions (8) demonstrates an order of magnitude advantage for photovoltaics over fossil fuels and allied technology. However, the integration of solar energy into a typical home system (11) shows that the use of lead-acid batteries dominates the life-cycle effects. Therefore the management of battery waste is a serious issue in this context, and the design process should consider increased battery life, improved battery specification and reduction/recycling principles. Although a number of battery manufacturers have attempted to improve battery life-cycle impacts, product developers have a responsibility to educate the end-customer in terms of battery management and proper recycling. Although the published LCA analysis is based on a whole range of assumptions, we can estimate that the greenhouse pay-back time is approximately doubled when there is an absence of battery recycling. In a similar study by Watt *et al.* (12) comparing stand-alone photovoltaic power generation with grid supply, the ability to reduce total energy use and substitute for fossil fuel-generated electricity had a significant impact on life-cycle emissions. It should be noted that there are inter- and intra-national discrepancies when analysing photovoltaic emissions (13), and there is little consistency in the available data due to unstructured data gathering; hence, LCA remains a time-intensive process.

The detailed shelter analysis in the previous section showed that the life-cycle impact of the aluminium material components should be considered due to the fact that they contribute a significant proportion of the shelter weight and raw material/manufacturing costs. A full LCA and inventory data for aluminium is provided by the International Aluminium Institute, but the life-cycle processes tend to be based on a primary aluminium ingot. A more practical study by Werner and Richter (14) has assessed the life-cycle impacts of aluminium window frames, which are similar in nature to the main extrusions employed in bus shelters. The study showed that there are no general disadvantages with using aluminium extrusions over other materials, where possible recycled aluminium content should be in the range 35-85%, and the aluminium components should be easily separated from other materials to avoid contamination on disassembly.

3.4 Solar array selection
The initial task was to consult with Cardiff Harbour Authority and discuss the important customer requirements in terms of internal lighting and display panels. It was agreed to remove the need for internal lighting in order to reduce energy use. This section will therefore review the selection and development of the photovoltaic system in order to provide a balance between performance, environmental impact and capital cost. There were a wide range of variables to consider (panels, battery specification, control system, display specification, etc.) for each possible configuration, but the design was narrowed down to three possible systems:

A.   Three solar panels, two batteries and 15-character LED display unit;
B.   Four solar panels, four batteries and 21-character LED display unit;
C.   Eight solar panels, seven batteries and 20-character LCD display unit.

Figure 2 summarises the power usage and associated peak power requirements for the three systems under review. The reduced potential daylight during the winter months increases the winter charging in order to maintain the five-day autonomy; this in turn governs the peak photovoltaic supply requirement. The LCD display (system C) draws the most energy, and explains the need for the seven batteries. This system was rejected due to its heightened environmental burden associated with battery mis-management. When it became clear that a 15-character LED display was adequate for the customer's needs, the twin-battery option (system A) was the clear solution. It should be noted that system A was the most economic solution, and conformed with the customer's request that for this trial project the solar-powered system should be as small as possible. System C would have covered the entire shelter roof area, and this would have been far too intrusive. In summary, the three solar panel/two battery option was chosen because it was cost effective, minimised design/manufacture/installation time, provided the required level of performance and had the lowest perceived environmental impact.

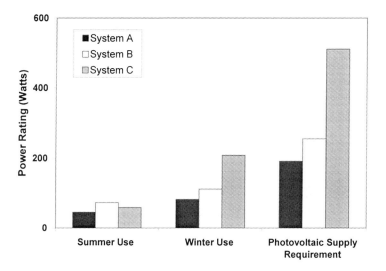

Figure 2. Comparison of power usage and nominal supply requirements for the photovoltaic systems

3.5 Prototype design and development
The main design complexity of incorporating solar panels into a standard shelter design is not associated with the novel photovoltaic array and charge controller, but rather with the effect of the increased weight due to the batteries. Although the minimum-battery system was selected, the main design challenge was to maintain the structural integrity of the roof sub-assembly without significantly increasing its weight or complexity.

The weight of a standard (non-solar) shelter roof is approximately 100kg; however, the new solar-powered roof would also need to accommodate 80kg of battery weight and 20kg of battery housings, as well as accommodating the weight of one person (80kg) for works access. This is a significant increase in roof weight, but detailed analysis of the RSJ roof beams highlighted a design solution that would compensate for this increase in battery weight. Standard roof beams are fabricated from rolled sections, which produced curved beams approximately 2m in length. Finite element analysis (FEA) of the nominal design showed that the factor of safety was in excess of 10 (typical analysis based on a uniformly distributed load is shown in Figure 3). The FEA analysis tool allowed the design team to access alternative design options, and it became clear that using a bespoke laser-cut beam would maintain the factor of safety but reduce the mass of the RSJ section. This was accomplished by using a thinner profile and castelating the flange section, thereby reducing the overall mass of the roof support beams. In addition to this weight compensation, the laser cut method also significantly reduced the proportion of scrap material produced. The rolled-beam technique would typically waste up to 25% of the original stock material, but the laser cut technique was far more efficient.

The final development task was to design the main body of the shelter. A modular glazing system was employed that could provide the structural integrity but that also conformed to the styling requirements of the solar-powered shelter. The prototype shelter design is shown in Figure 4 as a rendered 3D CAD image. It is estimated that the total design time was in the order of 100 hours. This was followed by the manufacturing and assembly phases, which took approximately four weeks.

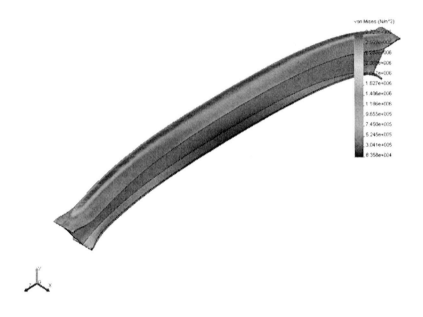

Figure 3. Finite element analysis of a standard curved roof beam.

Figure 4. Solar-powered shelter design

3.6 Customer feedback

Market research has followed the Cardiff Barrage project, and has targeted county councils with the aim of establishing the customer's longer-term sustainability needs. The preliminary feedback shows that customers are aware of Agenda 21, and realise that solar power is one mechanism by which councils can demonstrate a tangible commitment to the environment. Customers also need guidance on materials selection as they generally have little idea as to which materials are genuinely environmentally friendly. The immediate benefits of the solar-powered shelter concept from the customer's perspective cover time planning and cost savings. Connection to the national grid can take three months (i.e. longer than the shelter development time), and provision of the trench and electrical cabling costs approximately £60 per metre. These time and cost constraints are not incurred with a solar-powered shelter, saving in the region £600 to £1200 per installation. In extreme cases, councils need to cover higher electrical installation costs for remote rural locations, and a recent case in central London cost £15000 to cover electrical supply. Councils have also referenced the social needs of bus shelter development. The materials and construction need to be vandal resistant and provide features that deter crime. One application of solar power could be the power supply for safety lighting in conjunction with CCTV coverage.

4 DISCUSSION

The solar-powered bus shelter development described in this paper may be treated as a pilot project that has allowed a 'traditional' manufacturing company to start to incorporate eco-products and sustainable procedures within their business framework. The design results show that an SME can adapt and update standard products in order to satisfy the sustainability demands on key customers. However, the design and development reported here is iterative in nature, and conforms to the model of incremental ecodesign improvements rather than radical sustainable solutions. Long-term sustainability will need to harness the full range of ecodesign tools and techniques, and there is some question as to which are applicable to SMEs and which eco-design procedures require further development. In the context of SMEs, development projects are driven primarily by the issues of time and cost; consequently, the adoption of new procedures is judged in terms of how they impact on the bottom line. The time-intensive LCA procedures, which frequently generate ambiguous results, are therefore not appropriate for an SME in the early stages of a development programme. LCA output tends to be too general for specific day-to-day design problems, and the flexible nature of small-company low-batch-volume development projects means that more practical tools are needed. The main piece of design documentation that could form the basis for practical SME-friendly ecodesign procedures is the product design specification (PDS). The PDS needs to capture all the salient information, from materials and performance specifications through to packaging and shipping constraints. Therefore a 'green' PDS would appear to be a viable mechanism to drive the adoption of ecodesign procedures within SMEs.

The initial phase of market research, based on the solar-power shelter concept, demonstrates that more councils and local authorities are increasing the level of importance that they assign to sustainability issues. However, these key customers are not fully up to speed on eco-friendly materials and environmental management issues, therefore new products must address their immediate needs in the first instance, i.e. time planning and cost savings. The ability to link successful ecodesign initiates with tangible economic advantages therefore has a clear message for company and customer alike.

One of the main lessons arising from this work is that battery management and battery recycling considerations are key aspects of any solar-powered development. The design option with the minimum number of batteries was selected for this pilot development, but the twin-battery system still produced a significant increase in roof weight. Ordinarily, an increase in roof weight would generate increased structural mass within the associated support beams and columns. The engineering challenge for this project was to accommodate the new roof weight by adapting the supporting roof beams, and thereby compensate for the battery weight. This was achieved by using FEA tools to modify the roof beams in terms of employing a thinner castellated section, but the resulting safety factor was still well within requirements. Future work in this area could apply the FEA analysis tools to other key sub-assemblies, and further reduce the material burden whilst maintaining functionality. The preliminary materials analysis in this paper indicates that the aluminium components provide a prime opportunity for reducing the overall shelter weight, together with the associated raw material costs and manufacturing costs. An holistic approach to product development, underpinned by environmental and economic drivers, has many applications within the context of street furniture, therefore BSW's product portfolio should benefit significantly as a result.

 D004/009/2004

REFERENCES

1   **Berliner, C.** and **Brimson, J.** (1988) *Cost Management for Today's Advanced Manufacturing: The CAM-I Conceptual Design*, Harvard Business School Press, Boston.

2   **Holdway, R., Walker, D.** and **Hilton, M.** (2002) Eco-design and successful packaging, *Design Management Journal,* Vol. 13, pp. 45-53.

3   **Johansson, G.** (2002) Success factors for the integration of ecodesign in product development, *Environmental Management and Health*, Vol. 13, pp. 98-107.

4   **Vecalsteren, A.** (2001) Integrating the ecodesign concept in small and medium-sized enterprises, *Environmental Management and Health*, Vol. 12, pp. 347-355.

5   **Daly, M.** and **McCann, A.** (1990) How many small firms?, *Employment Gazette*, Vol. 100, pp. 14-19.

6   **Dahlstrom, H.** (1999) Company-specific guidelines, *Journal of Sustainable Product Design*, Vol. 8, pp. 18-24.

7   **Millward, H.** and **Lewis, A.** (2003) Product development limitations within SMEs: The drive to manufacture in the absence of an understanding of design, *Proceeding of the 10th International Product Development Management Conference*, Brussels, Belgium, pp. 713-723.

8   **Hamer, G.** (2004) Solar industry trends: Solar power key to national energy, *Solar Today*, Vol. 18, pp. 18-21.

9   **Kelly, T.** (2001) Prototyping is the shorthand of innovation, *Design Management Journal*, Vol. 12, pp. 35-42.

10 **Lipscomb, M.** and **McEwan, A.** (2001) The TCS model: An effective method of technology transfer at Kingston University, UK, *Industry and Higher Education,* December, pp. 393-401.

11 **Alsema, E.** (2000) Environmental life cycle assessment of solar home systems, Report NWS-E-2000-15, Department of Science Technology and Society, Utrecht University.

12 **Watt, M., Johnson, A., Ellis, M.** and **Outhred, H.** (1998) Life-cycle air emissions from PV power systems, *Progress in Photovoltaics: Research and Applications*, Vol. 6, pp. 127-136.

13 **Gurzenich, D.** (2003) Cumulative energy demand and cumulative emissions of photovoltaic systems in the European community, *International Journal of Life Cycle Assessment*, Vol. 8, p. 64.

14 **Werner, F.** and **Richter, K.** (2000) Economic allocation in LCA: A case study about aluminium window frames, *International Journal of Life Cycle Assessment*, Vol. 5, pp. 79-83.

'Closing the loop' – the role of design in the success of six small UK recycled product manufacturers

A CHICK and **P MICKLETHWAITE**
Faculty of Art, Design, and Music, Kingston University, UK

ABSTRACT

This paper introduces a multiple case study project which profiles six small UK enterprises (for–profit and not-for-profit) that manufacture recycled 'consumer durable' products. The challenges and opportunities faced by these enterprises are considered, along with their design approaches, and lessons for success are identified. The paper explores how recycled materials (recyclate) can be successfully incorporated into new product manufacture, providing the basis for new enterprise and firm competitiveness. Increased production of recycled products in the UK is a vital link in the cyclical production and consumption system expressed by the idea of 'closing the loop'.

1 RECYCLING: 'CLOSING THE LOOP'

There is currently a great deal of political emphasis in the UK on recycling as a strategy for sustainability (1). The UK Government has set aggressive recycling collection targets (Table 1).

Table 1. UK household waste recycling collection

At least 25%	UK Government target for household waste recycling [collecting] or composting by 2005 (1)
12.4%	The proportion of household waste collected in England in 2001/2002 (2)

There is little incentive to reach these targets without development of new commercial applications for the materials recovered from the waste stream. Increased recycling collection will only be achieved if it is made commercially viable by the reuse of collected recyclate in new manufacture and production. The remanufacture of recyclate into new products is key to 'closing the loop' (Figure 1).

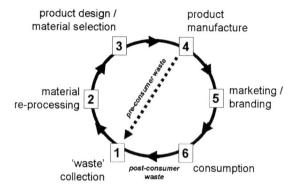

Figure 1. 'Closing the loop': cyclical production and consumption

This paper focuses on the role of design, in its various forms, in the six case studies. It complements the focus on recycled consumption, and its implications for design, in an accompanying paper submitted to this conference (3).

Production, marketing and consumption of recycled products is slowly growing in the UK.

"It remains important to build market demand for recycled products. In principle, if there is strong demand for a good or service, then individual entrepreneurs and organisations (both for-profit and not-for-profit) should be spurred into attempting to meet that demand." (4)

Small manufacturing enterprises have a vital role to play in developing markets for recyclate in the UK (5). Yet these companies are often less likely than larger companies to use recyclate, or consider other environmental issues (6, 7). UK designers are also vital to growing this sector. Yet they, similarly, don't generally consider environmental issues in their professional practice (8, 9, 10). It has been recognised that for the market for recycled products to increase, "enterprises have to be willing to re-assess their attitudes to product specification and to suppliers." (4) Furthermore, there are opportunities for enterprises to try novel approaches in terms of design which will improve their competitiveness. Current negative perceptions of recyclate and recycled products by designers and manufacturers need to be addressed.

Environmentally-considered material selection is one of the easiest ecodesign principles for small manufacturers to implement (6, 11). Introducing single issues such as using recyclate can then lead UK designers and manufacturers into deeper 'sustainable design' thinking and practice (12).

2 CASE STUDY SELECTION AND RESEARCH METHOD

A multiple case study approach was used to empirically investigate how six small UK enterprises (for–profit and not-for-profit) have successfully incorporated recyclate into competitive product design and manufacture. Six enterprises were selected using the qualification criteria shown in Table 2, which were developed to embody a deliberate research focus on diversion of waste from UK landfill for reprocessing and remanufacture in the UK (for further explanation of these criteria see the project website).

Table 2. Case study selection criteria

The six featured enterprises:
- are classified as 'small' (under 50 employees) or 'micro' (under 10 employees) (13)
- were established specifically to make recycled 'consumer durable' products *

The products they make are:
- high recycled content
- generally high post-consumer content
- made from recyclate recovered from a key UK material waste stream
- manufactured and sold in the UK
- sold in end-consumer markets
- explicitly marketed as 'recycled'
- currently, or have the potential to be, mass-produced

*Note: * "consumer goods ... whose useful life extends over a relatively long period [rather than those which are] used up within a short time after purchase." (14)*

Finding enterprises in the UK that match all these criteria proved a difficult and lengthy process. Candidates were found via preliminary internet searches and consultations with recycling organisations, then contacted and assessed for suitability by telephone.

Good product design or a design-led approach don't appear in the criteria. Ascribing labels like these is quite subjective, when a list of solely objective criteria was sought. Most of the featured enterprises in fact see themselves as 'design-led', and 'design' appears in the list of success factors identified below ('Lessons for Success'). Design is therefore identified as a key factor in developing attractive and successful recycled products, and a competitive enterprise.

An interview schedule was sent prior to a personal interview with the founder of each enterprise. Interviews were semi-structured; questions were consistent but responsive to specific interesting aspects in each case. All interviews were filmed by a professional production company, in order that interview clips could be used on a website created for the project described here (URL given below). It is possible that the paraphernalia and disruption caused by filming the interviews (rather than making simple written or audio recordings) had an intimidating effect on some interviewees. Greater efforts to put interviewees at ease were therefore required. The richness of the data collected in this way was greatly enhanced, however, allowing video clips to be used on the project website. A full audio transcription was made of each interview. This intimate engagement with the data allowed careful selection of the most relevant and interesting quotes for use in constructing the case studies. Case study profiles were built from various sources, including the interview data, informal telephone and email correspondence, and internal and external documentation such as websites and press coverage. This research approach is consistent with a general case study methodology incorporating a range of specific research methods (15). All research was conducted from mid-2003 to early 2004.

3 LESSONS FOR SUCCESS

The six enterprises differ widely in the materials they use, the products they manufacture, and the markets into which they sell. Yet there are some common factors in their success.

Market knowledge and product marketability: knowledge of the markets into which a product or product range is to be sold, and the marketability of that product or range, was the most prominent success factor across the six enterprises.

Product design, innovation and added-value: most of the featured enterprises see themselves as 'design-led', and identify 'design' as a key factor in developing attractive and successful recycled products.

Reliable supply of recyclates: to be useable in product manufacture, recyclates must meet standards of quality and be consistent in supply and price.

Production and process innovation: while product design and innovation are easily identifiable as sources of added value, production and process innovation can be equally key to the success of a new recycled product.

Determination to succeed: personal tenacity, persistence, and an unwillingness to give in are key entrepreneurial traits (16) displayed by the 'green' entrepreneurs featured here.

The case studies which follow examine 'tangible' design activity in the six featured enterprises, i.e., design activity which impacts on the physical properties of the product (17). The six case studies are presented in the following three 'pairs': Cutouts Ltd and Remarkable Pencils Ltd; The Green Bottle Unit and the Recycled Bottle Glass Centre; Re-Form furniture and Plastic Reclamation Ltd. These 'pairs' are defined by similarities in recyclates used, products manufactured and markets served. In addition, within each pair similar issues are faced, though these may be addressed in different ways. A table containing a summary profile of each case study enterprise is included at the end of the paper.

4 CUTOUTS LTD: SERVING A NICHE GREEN MARKET

Table 3. Cutouts Ltd

Established	1996
For-profit	Yes
Turnover	Unknown
Employees	4 full-time + seasonal
Products	Stationery and giftware
Recyclates	Printed circuit boards and various plastics
Annual volume of recyclates	Unknown
Product recycled content	High to 100%
Post-consumer waste content	Various

Cutouts is a small family-run company specialising in the manufacture and sale of recycled gift and stationery products (Table 3). Originally set up to produce toys for children, the company now produces a wide range of stationery and giftware items (including mouse mats, personal organisers, notebooks, clipboards, credit card holders, rulers and coasters) from reused Printed Circuit Boards and various recycled plastics. Cutouts products command a high price, and serve niche recycled giftware and stationery markets.

4.1 Design approach

Cutouts is a very small company, and does not have a designer on its staff. The company's Director nevertheless argues the case for good design as a source of product differentiation and added value.

> "Although our products are recycled it does not mean to say that they're going to sell just because they are recycled. They have to be of good design, they have to be attractive, and they have to be very competitive in the market that they're in, the gift market. ... They simply will not sell just because they are recycled, it is an added bonus."

New product design ideas are generated within the firm; commissioning external designers is considered a prohibitively expensive option, and one which does not provide any guarantee of generating suitable ideas for new Cutouts products. Thus, while the firm sees itself as being design-led, this is not manifest in a commitment to employing professional design expertise. Nevertheless, Cutouts' Director sees the company's products as having "a certain wow factor" deriving from their visual impact and uniqueness. Cutouts appears to be quite insular in much of its design activity, unwilling to risk commissioning external professional product designers. The company does outsource much of its production, however; of necessity, considering the broad range of materials and products it produces.

5 REMARKABLE PENCILS LTD: MOVING FROM NICHE TO MAINSTREAM MARKETS

Table 4. Remarkable Pencils Ltd

Established	1996
For-profit	Yes
Turnover	£1,000,000
Employees	15
Products	Stationery
Recyclates	Various plastics, rubber tyres, leather, paper
Annual volume of recyclates	50 tonnes of tyres; 4 million plastic cups; 100 tonnes of paper & card; 20 tonnes of computer printer materials
Product recycled content	High to 100%
Post-consumer waste content	High

Remarkable Pencils Ltd was set up to design and manufacture a pencil made from one recycled polystyrene vending cup (Table 4). The Remarkable pencil was brought to market following a two-year development process, and was a Millennium Product award winner. Remarkable now produces a wide range of retail and promotional stationery products, all made from recycled and sustainable sources, including mouse mats, pencil cases, rulers, notebooks, pens and coloured pencils. Retail stockists range from major supermarket chains to small independent shops. Standard products are also sold through educational and business catalogues. Promotional products are either personalised by Remarkable themselves, or sourced 'unbranded' by trade customers for their clients. Remarkable are seeking to break into mainstream retail markets by developing their brand.

5.1 Design approach

The welcome page to Remarkable's website proclaims the company's credentials as a design-led manufacturer of "fun, innovative and colourful" recycled stationery products (18). The Managing Director also makes the point,

> "we have taken really basic concepts of colour, attractiveness, quality and also the durability of the product and made it out of recyclate, which is something which has been slow in coming to this marketplace. ... Design is integral to the whole thing, it's not just the fact that it is recycled."

This echoes the quote from the Director of Cutouts above. The Managing Director of Remarkable expands on the company's approach to product design relating to the material source to the end product, "creating a symmetry between the waste material and the product itself, like for instance one cup makes one pencil. People have a tangible recognition with what the waste was to what the product is, and that … it's a form of design, and it's integral to how we move forward with the process and the products and also the image of the company."

In addition to product design and innovation, process innovation is also identified as adding value to Remarkable's products,

"it's not only the design of the products that we've introduced here, we actually designed the processing of the recycled material and the manufacturing of the products as well."

The success of the Remarkable recycled pencil, for example, manufactured from one recycled polystyrene vending cup, may be seen as being built upon the development of a process to convert the recovered source material into a form suitable for manufacture. The company is not wedded to the idea of in-house production, however, and appears keen to develop an external manufacturing network to support future expansion.

6 THE GREEN BOTTLE UNIT: 'THE ART OF RECYCLED GLASS'

Table 5. The Green Bottle Unit

Established	2000
For-profit	No
Turnover	£350,000
Employees	16 (11 full-time equivalent)
Products	Glass tiles, bricks and pavers
Recyclates	Glass
Annual volume of recyclates	*
Product recycled content	100%
Post-consumer waste content	100%

*Note: * GBU has a production target of 2000 tonnes of recycled glass per annum by 2006.*

The Green Bottle Unit (GBU) is part of a charity which brings together artists, designers and architects and excluded sections of the community in urban regeneration projects (Table 5). It was set up to demonstrate innovative new uses for recycled glass products in an urban environment. It produces high-specification 100% recycled glass tiles, bricks and pavers for internal and external use in a wide range of urban landscaping applications. GBU currently sells its products exclusively to order. The products currently command a high price due to the small scale of production, but GBU is ambitious to market its products via multiple Do-It-Yourself retailers.

6.1 Design approach
The design agenda of GBU is succinctly expressed in its promotional strapline: "the art of recycled glass". In the words of the Project Director,

"most people associate waste products as being rubbish, basically, and not of very good quality. So our mission is about the high quality of design through the waste of glass."

Recovered glass is kiln fired and finished to create products suitable for use in the design and construction industries. All products are currently handmade, and slight colour and size variations make each product aesthetically unique. GBU specialises in collaborating with designers on bespoke commissions, recently working with a leading international architectural practice for a Japanese client.

"within reason, anything that can be actually cast in glass, the images can be made [in recycled glass]. So any glass form that one can see out there could actually be made in recyclate."

Design, as conceived and practiced by GBU, is an activity performed with artistic credentials foremost in mind. Strategically, GBU is ambitious to market its products via multiple Do-It-Yourself retailers. This would require a massive increase in the capacity and degree of automation of production, however, posing significant financial and technical obstacles. Scaling up to mass production, and being beholden to large retailers, could also affect the artistically-oriented designer-maker approach thus far maintained by GBU. The Project Director talks about the need for products to have "a marketable edge, which makes people go 'I am prepared to pay for this as opposed to the alternatives.'" The dilemma faced by GBU is that this 'marketable edge' attracts interest and potential orders on a scale which it can't possibly currently deliver. An alternative solution to the issue of increasing the scale of production is franchising: "the unique designs could actually be sold on to other people to make." This idea is taken up in the next case study.

7 THE RECYCLED BOTTLE GLASS CENTRE: FRANCHISING RECYCLED GLASS MANUFACTURE

Table 6. The Recycled Bottle Glass Centre

Established	1997
For-profit	No
Turnover	£100,000 *
Employees	13 *
Products	Recycled glass sheets
Recyclates	Bottle glass
Annual volume of recyclates	600 tonnes *
Product recycled content	100%
Post-consumer waste content	100%

*Note: * These figures are for 2001/02, before RBGC switched from product manufacture to licensing its production process. Several licenses have been sold.*

The Recycled Bottle Glass Centre (RBGC) was set up to employ adults with disabilities in producing decorative products made from recycled glass bottles (Table 6). RBGC's recycled

'Phoenix Glass' has been used to refurbish and repair traditional stained glass windows. The company recently changed direction to concentrate on marketing licenses to use its patented production process to produce flat glass sheets from recycled bottles. It now only processes glass for use in stained glass projects and its demonstration and training unit. The shift to licensing of production in 2003 was done to meet growing demand for the product. Marketing of licenses is assisted by word-of-mouth recommendations and recycling market development organisations.

7.1 Design approach
RBGC was established by a stained glass artist, now its Technical Manager, and a craft sensibility is evident in his attachment to the material.

> "The qualities of the glass are the same as mouth-blown glass, which is very expensive. But the glass that we make is very cheap ..."

In its current craft-based applications, RBGC's recycled glass is technically superior to non-recycled alternatives. As a result, RBGC previously faced the same issue currently faced by the Green Bottle Unit, namely how to meet growing demand for its products in the face of considerable financial and technical obstacles to increasing the scale of production. The franchising option discussed by GBU has been taken forward by RBGC with some initial success; a number of licenses have been sold within the UK and enquiries received from outside the UK.

> "what we wanted was for other people that have got different ideas to come up with different ways of using it [Phoenix Glass] ... it's an open license, people can buy the expertise based on the patent so it's protected, and they can develop whatever they want and sell whatever they want wherever they want ..."

How the material is used is therefore delegated to the licensees, which opens up a much broader range of potential applications and design possibilities than would have otherwise been the case.

8 RE-FORM FURNITURE: RECYCLATE AND THE DESIGNER-MAKER

Table 7. Re-Form furniture

Established	1992
For-profit	Yes
Turnover	Unknown
Employees	1
Products	Furniture
Recyclates	Timber and plastic
Annual volume of recyclates	Unknown
Product recycled content	100%
Post-consumer waste content	100%

The designer-maker behind Re-Form furniture established a workshop to design and make furniture in 1992. Re-Form furniture is made from reclaimed timber and recycled plastic sheets made from post-consumer waste polyethylene bottles. The Re-Form range includes chairs, stools and tables which exploit the durability and unique aesthetics of the recyclate used. All Re-Form products are currently handmade to order and so command a relatively high customer price. The designer is now looking to implement a business plan which will scale-up and automate production, making the Re-Form range more competitive.

8.1 Design approach

Re-Form furniture is the product of its designer-maker's twin desire to address the issue of waste and explore the craft and production possibilities of recyclate. The designer extols the virtues of the recyclate he works with, making it clear that his personal use of them is very much determined by the qualities of the materials themselves. Re-Form furniture consciously exploits the striking visual aesthetic of the recycled plastic sheeting used in its construction.

"every piece is unique because no piece of plastic is the same. Every batch of plastic differs in terms of its colour and in terms of how big the spots or the gradients of the colour are. So that's always quite interesting to see how a new sheet is gonna look on a piece of furniture."

Re-Form furniture is additionally designed for easy dis-assembly and re-assembly, enabling 'flat pack' transportation or storage. The twin issues of cost and scale of production arise here as elsewhere in the case studies. Unit cost is currently high, yet can only decrease with investment in the automation and scale of production, which may in turn impact on aspects of the products' design.

9 PLASTIC RECLAMATION LTD: HIGH VOLUME RECYCLED PRODUCT MANUFACTURE

Table 8. Plastic Reclamation Ltd

Established	mid-1970s
For-profit	Yes
Turnover	£2,000,000
Employees	40
Products	Highway, outdoor and street furniture; landscaping and waterways products
Recyclates	Plastics
Annual volume of recyclates	10,000 tonnes
Product recycled content	100%
Post-consumer waste content	90%

Plastic Reclamation Ltd (PRL) has been involved in plastics recycling and product development since the mid-1970s (Table 8). The company manufactures and markets a wide range of 100% recycled plastic ('Epoch') and plastic and wood composite ('Knotwood') highway, outdoor and street furniture, and landscaping and waterways products. PRL markets its products to a diverse range of customers, including local authorities, industrial customers and the general public. It also offers commission-moulding services, and is an Accredited Plastic Reprocessor.

9.1 Design approach

The PRL product range is built upon material substitution; conventional products are manufactured using recyclate rather than conventional non-recycled alternatives. One of the recyclates developed by the company is marketed as Knotwood, in response to the apparently common reaction 'that's not wood'. The primary design activity is therefore process innovation, rather than original product design. The recyclates developed by PRL have qualities which make them superior to non-recycled alternatives in the applications for which they are used. PRL's engagement with design is at present quite limited. The company recognises that more innovative product design could allow it to develop new product opportunities, and thereby increase the appeal of its product range to wider consumer markets.

10 CONCLUSION

The six case studies examine design activities which impact at 'product' and 'process' levels (19). The design activities prominently discussed are:

- Product design, including material substitution
- Manufacturing process design
- Material development
- Outsourcing production

The majority of the cases have successfully used design at the 'product level'. All the enterprises realise that they need to establish a reputation for quality products as a basis for competitiveness and expansion. This objective is emphasised by the negative perceptions of recycled products that predominate in the UK (4). At the 'process level' a number of the enterprises would benefit greatly from focusing on designing for manufacturing. Technical assistance to overcome the various production obstacles which currently hinder any increase in the scale and speed of production would also be valuable, notably in the case of the Green Bottle Unit. Process innovation barriers are not uncommon in the recycled product manufacturing sector as this is often uncharted territory in the UK (4). Those enterprises encountering process level barriers will need to harness and mobilise their product and production capabilities before they can truly develop the 'intangible' design activities discussed below.

It is limiting to focus on design purely in terms of the tangible aspects of the product. A "narrow engineering-based view of design [which] neglect[s] the importance of other intangible aspects of the design process" is incapable of accounting for "other more subtle/intangible ways of distinguishing their products from their competitors." (17) There is thus a much broader range of 'intangible' design-related activities, which "can [be] viewed at a *strategic level* in terms of brand building at one end of the spectrum and the development of a corporate design culture at the other." (19) These activities go beyond what could be considered as relating directly to product design.

Design, in this broader sense, is a common thread running through the 'Lessons for Success' derived from the case studies above. The six enterprises demonstrate a broader range of design activity than that originally investigated in the case studies above. Prominent additional intangible design activities relate broadly to the enterprises' communication efforts, and include marketing, branding and website design. The six enterprises engage in these activities to differing extents and in differing combinations, and claims to be truly design-led in a broader sense appear to be much stronger in some cases than in others. Remarkable Pencils Ltd is the exemplar here in this regard, insofar as it demonstrates the strongest commitment to design in embracing both tangible and intangible design activities. The other five enterprises are more limited in their appreciation and use of design; they also recognise their limitations to varying degrees. Cutouts, for example, seems keen to emulate Remarkable's attempt to move out of a 'green niche' and into mainstream retail stationery markets by the use of branding. Yet Cutouts remains extremely wary of bringing in professional design assistance, and accepting the financial speculation and risk, which such a step would involve. This view appears to reflect that of SMEs generally towards brand development and seeking assistance from brand and design consultancies (20).

The authors conclude that the six featured enterprises would greatly benefit from developing a more holistic approach to design, "seeing it as a set of interrelated processes that cut across many different parts of the firms' activities, such as engineering and industrial design, branding and marketing." (17) What is important here is that the enterprises develop both existing and new sets of design-related capabilities that interrelate with each other and allow them to compete profitably, with quality products at competitive prices and genuine consumer appeal.

11 NOTE

The six case studies are profiled in more detail at www.inspirerecycle.org, the website for which the research described in this paper was undertaken.

12 ACKNOWLEDGEMENT

This project has been developed within Phase II of the Recycling by Design Research Project (www.recyclingbydesign.org.uk), funded by Biffaward, London Remade and LRL Ltd.

REFERENCES

1. **DETR** (Department of the Environment, Transport and the Regions) (2000) Waste Strategy 2000 for England and Wales. http://www.defra.gov.uk

2. **DEFRA** (Department for Environment, Food & Rural Affairs) (2003) Municipal waste management statistics 2001/02 (16 September). http://www.defra.gov.uk

3. **Micklethwaite, P.** (2004) The 'recycled consumer': evidence and design implications. Paper submitted to Design and Manufacture for Sustainable Development Conference, Loughborough, 2004.

4. **LDA** (London Development Agency) (2003) Green Alchemy – Turning Green to Gold: Creating resources from London's waste. London: LDA.

5. **Watts, B. M., Probert, J.** and **Bentley, S. P.** (2001) Developing markets for recyclate: perspectives from south Wales. Resources, Conservation & Recycling, 32, pp 293-304.

6. **Smith, M. T., Roy, R.** and **Potter, S.** (1995) The Commercial Impacts of Green Product Development. Milton Keynes: Open University.

7. **de Graaf, D.** (2002) Development of a tool to introduce product oriented environmental management at SMEs. In Charter, M. (ed) Managing Sustainable Products: Organisational aspects of product and service development. Towards Sustainable Product Design 7 Conference Proceedings. Farnham, Surrey: The Centre for Sustainable Design.

8. **Dewberry, E.** (1996) E codesign - Present Attitudes and Future Directions. PhD thesis: Open University.

9. **Chick, A.** and **Sherwin, C.** (1997) 'Eco-design' Survey: a study of the Chartered Society of Designers (CSD) membership. Farnham, Surrey: The Centre for Sustainable Design.

10. **Chick, A.** and **Micklethwaite, P.** (2004) Specifying recycled: understanding UK architects' and designers' practices and experience. Design Studies, 25 (3), pp 251-273.

11. **van Hemel, C.** and **Cramer, J.** (2002) Barriers and stimuli for ecodesign in SMEs. Journal of Cleaner Production, 10, pp 439-453.

12. **Sherwin, C.** and **Bhamra, T.** (2000) Innovative ecodesign: an exploratory study. The Design Journal, 3 (3), pp 45-56.

13. **SBS** (Small Business Service) (2004) Research and Statistics. http://www.sbs.gov.uk

14. **Pallister, J.** and **Isaacs, A.** ed. (1996) A Dictionary of Business. 2nd edition. Oxford: OUP.

15. **Gillham, B.** (2000) Case Study Research Methods. Real World Research series. London: Continuum.

16. **Muzyka, D. F.** (1998) Entrepreneurship: An overview. Presentation made at Enterprising Europe Conference, 6-7 April 1998, London.

17. **Whyte, J. K., Davies, A., Salter, A.J.** and **Gann, D.M.** (2003) Designing to compete: lessons from Millennium Product winners. Design Studies, September, pp 395-409.

18. **Remarkable** (2004) Homepage. http://www.remarkable.co.uk

19. **Trueman, M.** and **Jobber, D.** (1998) Competing through design. Long Range Planning, 31 (4), pp 594-605.

20. **Colyer, E.** (2002) Can small businesses sprout big brands? http://www.interbrand.com [Published 09 December]

.

Authors' Index